Plants and Insects Together

Plants and Insects Together

by Dorothy Hinshaw Patent

drawings by Matthew Kalmenoff

HOLIDAY HOUSE · NEW YORK

TITLE PAGE DRAWING: *Nectar guides—lines radiating from the center of a flower—may be visible or invisible, but either way they help an insect to find nectar. See Chapter 3.*

LIBRARY OF CONGRESS CATALOGING IN PUBLICATION DATA

Patent, Dorothy Hinshaw.
  Plants and insects together.

  Bibliography: p. 123.
  Includes index.
  SUMMARY: Describes the many varied relationships
between plants and insects.
    1. Insect-plant relationships—Juvenile literature.
[1. Insect-plant relationships] I. Kalmenoff, Matthew.
II. Title.
QL467.2.P37   595.7'05'24   75-34205
ISBN 0-8234-0274-6

TO DAD,
who always encouraged my scientific interests

Contents

1

Plants and Insects:
Partners/Enemies

Plants and insects, insects and plants—both are very familiar to us, and we often see them together. But have you ever wondered how insects and plants relate to one another? In some ways they are partners; in other ways they are enemies. Both groups have been on earth for a long time and have developed many fascinating relationships.

We know from studying evolution how the different kinds of living things on earth slowly developed over millions of years. It is very hard to imagine how all the great variety of life as it exists today could have had its start in simple, single-celled organisms. But it has been about three billion years since the first living things appeared on earth. A lot can happen in three billion years.

For a long time after life began, all living things lived in the sea. But hundreds of millions of years ago, life began to move out onto the land. Plants and insects were among the first to leave the water. They have since evolved into a bewildering variety of life. And they have evolved together, affecting one another in many ways. Among the plants and insects alive today, we see the results of those long years of evolving together, or coevolution. Some insects help plants, and some plants help insects. Other insects eat plants, and some plants even eat insects. Some of these relationships are very complicated.

Camouflage

Many insects disguise themselves—in effect—by resembling parts of plants. They blend so well into their surroundings that enemies have trouble finding them. Over the ages, insects which resembled plant parts were less likely to be eaten before reproducing. Therefore disguised insects produced more offspring, and many beautifully camouflaged insects are found today.

The wings of many moths have patterns in brown shades which blend perfectly with tree bark. The underside of some butterfly wings have brown patterns.

Right: *Some ways that insects "get lost" on plants: at upper left, a moth's wings blend with the bark of a birch; at upper right, a butterfly with wings remarkably resembling leaves. Lower left, a species of caterpillar (to the right of the leaves) that can barely be distinguished from a twig; and the walkingstick, which also looks like woody twigs.*

With their wings folded, such butterflies look for all the world like dead leaves. The wings of some kinds have markings which look just like the veins of leaves. They have notches around the edges, as if the leaf had been nibbled on. Caterpillars are often the same shade of green as the leaves they eat and are therefore hard to find. Other caterpillars resemble little twigs.

The stick and leaf insects look amazingly like the plants they live and feed on. Stick insects have long, thin bodies which have bumps and "thorns" just like those of their food plants. Leaf insects have flattened bodies. Their wings and legs have thin outgrowths with the form and color of their host plant, so they are almost invisible. The eggs of leaf and twig insects often look just like the seeds of the plants they feed on.

Some insects are so camouflaged that they can hide from their prey. The praying mantis is green and sits motionless on a twig, waiting for passing insects. When one gets close enough, the mantis grabs it in a flash. Some tropical mantises are disguised as flowers. One kind looks so much like the red orchids upon which it sits that unsuspecting insects probe it for nectar before being snapped up. Another tropical mantis hangs down from a plant. It has green, red and white markings which make it look just like a flower. It feeds on the

Left: *The common greenish praying mantis sits still, not very noticeable, until it can grab an insect that comes near. Some tropical mantises resemble flowers* (lower left) *or leaves.*

flies and butterflies which mistakenly approach it in their search for nectar.

Some weevils which live in the high, remote moss forests of New Guinea camouflage themselves in a totally different way. While most weevils have smooth backs, these large and flightless insects have strange scales and hairs on their backs, as well as pits and knobs. The roughness thus produced plus a special waxy secretion make the weevil's back an ideal home for several kinds of small plants. Mosses, algae, lichens, and other plants grow there. The miniature jungle even houses a unique mite and several other small animals. These weevils are very slow-moving, yet seem rarely to be eaten. Whether this is because of actual camouflage or because of the bad taste of their portable gardens is unknown.

Plant Defenses

Many scientists study the relationships between insects and the plants they eat. Common insects such as grasshoppers, caterpillars, aphids, and beetles feed on plants, yet plants are able to survive and flourish. Why haven't all the plants been gobbled up long ago? Plants have developed, through evolution, ways of protecting themselves from insects. Insects in turn have evolved ways around some plant defenses. There is a constant "evolutionary race" between insects and the plants they feed on. A change in the plant may lead to an adaptive change in the insect which in turn results in another change in the plant, and so forth. The evolutionary race is going on even today and will continue as long as plants and insects inhabit the earth.

Trees in particular would seem to be easy targets for insect attack. They are large and do not die off every year as many plants do. But trees are far from helpless. They are protected in several ways from insect attack. The dry, outer bark of many trees makes it very difficult for insects to reach the moist living part of the tree. Bark often contains poisons as well. Many trees produce poisonous, sticky resins which stop the boring attempts of insects and ward off wound infections.

The leaves of trees are often almost inedible. They frequently have tough, slippery surfaces. Holly and oak leaves have spiny edges, too, which make it very hard for insects to hang on, much less chew. Many trees have poisonous leaves which could kill an insect which tried to eat them. The food value and water content of tree leaves is often low, so even an insect which could eat them would not grow very well.

Many insects prey upon the fruits or seeds of trees and other plants. But some seeds are poisonous and kill any creature which tries to eat them. Another way plants protect themselves from fruit- or seed-eaters is by bearing a crop every other year. The insects must leave their host each year in search of another which will be fruiting. In this way, the population of insects in and around any one tree can never build up too high.

Smaller plants sometimes use different defenses than trees do, although many of them have tough or poisonous stems and leaves as well. Many plants grow very fast and make seeds early so that they have reproduced before insects can destroy them. Plants which grow in fields mixed up with many other kinds of plants are often hard for insects to locate.

Hairs and Spines

Trichomes, which are very small hairs on plant leaves or stems, are a very effective defense against insect attack. Many plants have trichomes of different kinds. They ward off insects in a variety of ways. Some varieties of wheat are resistant to the cereal leaf beetle because of their trichomes. The female beetles are not likely to lay eggs on the hairy leaves. Even if the eggs are laid, they dry out and die easily. If some eggs should hatch, the dense hairs make it difficult for the larvae to get at the leaves to eat.

The trichomes of a tropical vine are very good caterpillar killers. While its relatives are being badly chewed up, this vine remains untouched. Every part of the vine, even the curling tendrils, is covered with sharp, hooked trichomes. Any caterpillar foolish enough to venture onto this vine soon has each of its fleshy legs caught by the hooks. The little wounds bleed, and the trapped caterpillar cannot move around to eat. Soon it dies.

Some scientists thought that plants which have such sharp trichomes had developed a defense which insects could not get around. But an insect has been found which has a very simple way of protecting itself from the large, spiny trichomes of its food plant. These are the caterpillars of a certain butterfly, and they live in groups of four to six. They spin a silken rug beneath themselves, always remaining on the underside of the leaves. Thus they can hang onto the silk while gravity keeps their bodies pulled downward, away from the sharp spines.

The most interesting trichomes are ones with glands. Some of these produce chemicals which inhibit feeding. Tobacco trichomes secrete poisons which paralyze the legs of aphids and kill them. The leaves and stems of wild potatoes are thickly covered with trichomes. If one is touched by an aphid, it releases a clear liquid. On contact with the oxygen in the air, this liquid is transformed into a black, insoluble substance which coats the aphid's legs. As the aphid explores the plant, it triggers more and more trichomes until it is so covered with the black material that it cannot move. Eventually it starves to death.

Chemical Defenses

Many plants without trichomes also use chemicals to defend themselves from insects and other hungry animals. Plants often contain chemicals called "secondary substances" which are not necessary for the nutrition of the plant. For a long time scientists argued about why these chemicals were present. Then, in 1959, Dr. G. Fraenkel suggested that secondary substances protected plants from attack by insects. He also said that some insects had developed a protective immunity to these chemicals. His ideas have interested many biologists studying these relationships, and now we know many new facts about these chemicals.

Quite a number of common vegetable plants, such as potatoes, tomatoes, cucumbers, squash, cabbage, and broccoli, contain various secondary substances. Some are powerful poisons, such as those of the castor bean or oleander leaves. Others repel insects. The active chemical in catnip is one example. Insects avoid touching an object coated with it and move away. Some protective chemicals keep the insects from biting the leaf at all. Still others may affect the growth of any insect which does feed on the leaves. One kind of plant may contain several different chemicals of this kind too, so that it resists insect attack in more than one way.

Insects Choose Their Food

By now we may wonder how insects can ever eat plants and thrive. Plants are not their typical food; more kinds of insects eat animal food than plants. Those insects which do feed successfully on plants have had to overcome many obstacles. Life on a plant leaf is considerably drier than it is down on the moist, shaded ground. These insects have to be resistant to drying out. Just hanging onto a plant can be a problem, especially when the wind blows. Some insects use silk to hang on, while caterpillars have short, fat legs equipped with suckers and little hooks. Some insects stay on leaf edges where they can get a good grip, while others bore inside the leaf or stem rather than remain outside.

Have you ever caught a caterpillar and tried to feed it leaves it would not eat? Some insects are very particular and will eat only one kind of plant. Others will eat a few kinds, while still others eat many different plants. There are advantages to each of these feeding strategies.

An insect which eats many kinds of plants has little trouble finding food. As long as a variety of food is available it will probably be healthy. But some plants are less nutritious than others. These insects could suffer if they fed too long on plants without much protein and water in them. They also must be able to recognize poisonous plants and avoid them if necessary. Some insects which feed on many different plants can stand a wide variety of poisons. They are even harder to kill with man-made pesticides than other insects are.

Some of the most successful insects are those which feed on a few kinds of plants. The monarch butterfly and swallowtail butterflies are familiar examples. Many crop pests, such as the tobacco hornworm, Colorado potato beetle, and cabbage butterfly are also in this group. These insects have turned the tables on the plants. They use plant secondary substances as telltale clues to locate their food plants. They are attracted by them rather than repelled. This is possible because they are resistant to the poisons and deterrents of their

The Colorado potato beetle is one insect that feeds mostly on the plants of the white potato. The potatoes' "secondary substances," chemicals that usually defend a plant against insect attack, are an attractant to this beetle, which damages potato crops heavily.

own food plants. They gain an extra advantage if cows, deer, and other grazing animals avoid their food, for they do not run the risk of being eaten along with the plant. Also, when an insect is limited to a few food plants or to only one plant, it can develop color and behavior patterns which hide it very effectively.

Caterpillars

Caterpillars are among the most common insects which limit themselves to a few food plants. If a caterpillar is going to eat the right food, first the female butterfly or moth must lay her eggs in the right place. Different aspects of plants can influence egg-laying, such as the surface texture of the leaves, or the amount of moisture in them.

The female European cabbage butterfly is attracted to the colors green and bluish green when she is ready to lay eggs. After she lands on a leaf, she drums her forelegs against the leaf, apparently sensing the chemicals in it. She has special sense cells on her legs, located in small hairs. These cells are sensitive to certain cabbage family secondary substances. These chemicals, especially one called "sinigrin," stimulate her to lay eggs on the leaf. If the sense cells are put out of action by a scientist, the butterfly lays no eggs at all.

Cabbageworms, diamondback moth caterpillars, several kinds of beetles, and a kind of aphid all specialize in feeding on cabbage family members. They are

stimulated to feed by sinigrin or other secondary substances of the plants. They are not injured by these
chemicals, for their bodies have ways of neutralizing
the poisons. Scientists have shown that other insects
may die from eating sinigrin, however. If celery, which
is a natural food of the black swallowtail butterfly caterpillar, is cultured in water containing sinigrin, the
chemical is carried into the leaves. Swallowtail caterpillars will feed on the leaves anyway; apparently they
lack sense cells which can detect this particular poison.
In nature they do not need such a warning system,
but in the laboratory they die if they eat celery leaves
containing about the same amount of sinigrin found in
some cabbage family plants.

The interaction between attractants and repellents
can be very complicated. If sinigrin is spread on leaves
of some unacceptable plants, cabbage worms will eat
them. But if leaves of poisonous potato family plants
are used, the caterpillars will refuse to eat. Other insects may eat only their own food plant, no matter how
much other leaves are doctored up.

Hormone Mimics

Some plants may have an especially effective way of
fending off hungry insects. Recently, many plants have
been found to contain chemicals similar to insect hormones. One such discovery came about in a very strange
way. In 1964 a scientist named Karel Slàma came to the

United States from Czechoslovakia to work with Dr. Carroll Williams of Harvard University. He brought with him cultures of the red linden bug, an insect he was studying. The bug had grown well in the laboratory in Czechoslovakia, but at Harvard the insects did not mature. Linden bugs have what is called "incomplete metamorphosis." Instead of having a larva which turns into a pupa and then emerges as an adult, the immature bugs look similar to adults. However, they lack wings and cannot reproduce. The larvae molt as they grow, with the fifth larval stage molting into the adult. The insects Slàma tried to raise at Harvard, however, molted into a strange sixth larval stage and sometimes even into a seventh! All the insects died without becoming adults.

This was all very puzzling and frustrating to Slàma and Williams, so they studied all the differences in the way they were raising the insects. Finally they focused on the different paper used in cultures for the bugs to walk on. In Prague, Slàma had used bits of filter paper. At Harvard he substituted paper towels. When he tried using filter paper again at Harvard, the bugs grew normally.

Thus began the strange search for the chemical dubbed "paper factor." Twenty different kinds of paper were tested on the linden bugs, including paper towels, toilet paper, and napkins. Twelve of them affected bug development. Then newspapers and magazines were tested. Strangely enough, all American news-

papers and magazines tested affected the bugs, while
most Japanese and European ones did not.

The scientists then performed chemical extractions
on the original kind of paper towels. They found that
they contained a chemical very similar to the insect
juvenile hormone. Juvenile hormone keeps immature
insects from turning into adults, and the related com-
pound from paper was having the same effect. The
chemical, which is called "juvabione," was eventually
traced to the balsam fir which is often used for making
paper in the United States. Other trees, such as tama-
rack and eastern hemlock, were also found to produce it.

Juvabione is active only against insects of the linden
bug family. But other plants contain different juvenile
hormone mimics which are effective against different
insects. Many plants contain chemicals which affect
insects somewhat as other insect hormones do. The
"favorite" one which plants mimic is the molting hormone. Various plants produce different such mimics;
some of these may kill insects by interfering with normal molting, while others appear to act as powerful
feeding deterrents.

Ways Around Tree Defenses

Some insects have even found ways around the defenses of trees. There are many types of bark beetles
which attack different kinds of trees. Each kind of
beetle specializes in feeding on one kind of tree or at

most a few kinds. The resins of nonfood trees are poisonous to the beetles, but they are attracted by the resins of their own food tree. Thousands of beetles will bore into one tree at the same time, killing the tree or injuring it so badly it cannot defend itself any longer. Some bark beetles carry with them a fungus which affects the tree. The fungus helps weaken tree defenses and may also be fed upon by the beetle larvae.

Sawflies successfully attack pine trees. The female sawfly has a sawlike egg-laying structure at the tip of her abdomen. With it she can saw into tough pine needles and lay her eggs there. The caterpillars are resistant to the poisons of the trees they feed on. The caterpillars of most successful kinds live in groups. Only some of them are strong enough to break through the tough surface of the pine needles to feed. After these stronger caterpillars have opened the needle, the others join in and feed successfully.

Some sawflies attack especially poisonous trees. The poison lies mainly in the core of the needles. Sawfly larvae which feed on these trees "skeletonize" the needles, eating only the safe outer part. If given needles of other fir trees to eat, they use the same feeding pattern, even though the cores are not poisonous. Larvae which normally feed on other kinds will die if they are fed the poisonous type, because they go right ahead and eat the whole needle.

One kind of oak tree has a very effective defense against many insects. Its leaves contain chemicals

which combine with proteins, making them almost indigestible. Insects feeding on these leaves would starve. But the eggs of the winter moth hatch especially early in the spring and the larvae can feed on young leaves which have not yet built up a high poison content. The oak is still ahead in the evolutionary race, however, for most winter moth eggs hatch too early and the larvae starve. Only those from eggs laid on trees whose leaves come out early survive.

Food Shortages

In most parts of the world there is a yearly season when food is scarce. Temperate regions have winter, while many tropical places have a long dry season. Since most plants lose their leaves at these times, leaf-eating animals must somehow survive with little or no food. Insects solve the food shortage problem in various ways. Some larvae simply stop growing as they molt, while some may actually become smaller. The most amazing case of such shrinking is that of the beetle larva Trogoderma. Some of these were kept without food for five years. At the end of that time they were still alive, but had molted down to one six-hundredth of their original weight. If there is only enough

Left: *A close-up of the red-headed pine sawfly and its work. Larvae are seen feeding at top, with an individual mature larva below. At left, a female deposits egg masses, shown in detail in a pine needle.*

food to stay alive without growing, larvae may stay the same size for years. The young of one type of wood-boring beetle have stayed alive without maturing in dry wood for forty years or more.

The most common solution to food shortages, however, is for the insect to enter a state called "diapause," which is a kind of suspended animation. Growth ceases and metabolism—the chemical process of turning food into energy—continues at a rate just sufficient for the insect to survive. In this way, food stores in the body are used up very slowly.

Insects undergo diapause at various stages of their life cycle. Scientists have investigated what it is that signals an insect to begin and end diapause, especially when overwintering. Most insects begin diapause in response to the decreasing length of the days. For example, the European cabbage butterfly may produce two or three generations during the summer. As long as there are more than thirteen hours of daylight, the larvae and pupae continue normal development. But when the day becomes shorter than thirteen hours, the caterpillars produce pupae which do not develop until the following spring.

Most insects which spend the winter in diapause require several weeks or months of cold weather before they can resume growth. Silkworm eggs, for example, will never hatch if they are kept warm all winter; they need two or three months of cool temperatures.

2
Chemicals on Loan

When insects eat plants, they consume a great variety of chemicals. Much of the plant material passes through the insect unused. Some of it is digested and provides energy for the animal. Other plant chemicals, such as vitamins, are needed to maintain health. Proteins are needed for their growth. And some insects have taken advantage of the poisonous secondary substances present in plants by storing them up for their own use.

As we have seen, some insects are immune in one way or another to the poisons of certain plants. Different kinds of insects deal with plant poisons in varying ways. Some merely pass them through their digestive tracts unchanged, never absorbing them into their bodies. Others have enzymes which convert poisons into harmless chemicals. Such insects gain special benefits from their deadly diet. They avoid competition with other plant-eaters, since only a limited number of insects are able to feed on poisonous plants. They avoid being eaten by grazers if their food plants are poisonous to mammals as well. They may also benefit if the other plants around theirs have been eaten. Then their leafy home becomes an isolated "island" not touching other plants. This makes it more difficult for insect predators and parasites to get to it.

The most successful insects which feed on poisonous plants, however, are those which are able to store the poisons in their own bodies. Some of the most common and widespread insects in the world, such as the monarch butterfly, belong in this category. The monarch belongs to a remarkable family of butterflies, the Danaidae. All danaids feed as caterpillars on plants of the milkweed family. Many milkweeds contain powerful heart poisons which are related to the drug digitalis.

These poisons make vertebrate hearts beat more strongly and slowly. They can kill, but a smaller-than-deadly dose makes an animal violently nauseated. This is a very unpleasant experience, but it makes the body rid itself of the poison. The monarch caterpillars store these poisons in their bodies and pass them along to the pupa and adult. One adult or pupa may contain more than enough heart poison to kill a cat. If an inexperienced bird eats a monarch, the mistake is soon apparent. Shortly after eating the butterfly, the bird vomits violently, suffering great pain and distress. This is an experience to remember. Once a bird has tried to eat a monarch it will avoid monarchs from then on, unless it is desperately hungry.

Bird predators have good eyesight and excellent memories, and most poisonous insects are strikingly patterned and colored, thus easy to remember. This helps the birds to avoid making the same mistake again, saving the poisonous insects from being attacked.

Not all monarchs are poisonous. It has been found that anywhere from 10 to 90 per cent of the monarchs in one location are poisonous enough to make a bird vomit. Only two out of the eight milkweed species in North America produce butterflies that poisonous. But even if just 25 per cent of the monarchs are that poisonous, the insects will be 75 per cent immune to bird attacks. This is because once a bird has eaten just one poisonous monarch it will eat no more.

Often other perfectly harmless insects will, so to speak, take advantage of the warning coloration and mimic the appearance of the poisonous insect. The viceroy butterfly is a good example. It looks very much like the poisonous monarch and is thus left alone by birds which have tried to eat monarchs. There are many examples of such mimicry in the insect world, particularly among tropical butterflies. In many cases, several types of poisonous butterflies have come to resemble one another. Each species then gains protection from the unpalatability of all the butterflies which it resembles.

Other insects can store poisons from plants, too. Moths, beetles, plant bugs, flies, lacewings, aphids, a scale insect, and grasshoppers which store poisons have all been studied. All such insects investigated so far are warningly colored as if to say "danger, stay away."

The problem of poisonous resins in pines is solved by one kind of sawfly in a very effective way. The larvae have two large pouches which branch off from the beginning of the digestive tract. The whole region is lined with protective cuticle, a kind of sturdy membrane or "skin." In some way, poisonous resins are extracted from the pine needles before they are swallowed. They are stored in the two sacs.

If an enemy attacks a larva, it rears and releases a droplet of the stored resin from its mouth. If any part of its body is poked or pinched, it turns around and dabs the fluid on its enemy. The larva is very quick and accurate in its aim. For example, it can reach any part of its own body except the back of its head.

The fluid is effective protection against ants and spiders. An ant which has been dabbed will stop attacking and carefully clean its body. By storing the resins, the sawfly larva gains their protection. At the same time, it avoids having to digest them or turn them into a harmless form.

Other insects feeding on poisonous plants may not wait until an attacker bites to show their unpleasantness. One grasshopper which feeds on repellent plants will regurgitate its food if attacked by ants. The ants then leave the grasshopper alone. Many insects use defensive sprays or secretions to protect themselves. Although some of these are known to be made in the insect's body, others probably are taken from food plants.

Sex Attractants

Female insects of many kinds attract mates with chemicals released into the air. These chemicals are called sex attractants. When the male senses the attractant with his antennae, he follows the odor upwind until he locates the female. Until recently, the scientists studying insect sex attractants have assumed that the chemicals are made inside the insect's body and are always the same for one species. But it now appears that at least some insects store plant chemicals and later use them as sex attractants. Females of the same species may have differing attractant mixtures, depending on their larval food plant.

The oak leaf roller moth causes a great deal of damage to oaks in the northeastern United States. In Pennsylvania a million acres have been infested by this pest in recent years, killing as many as 90 per cent of the trees in some areas. A group of scientists there is studying this insect and its sex attractants in the hope of finding ways to combat it without using poisonous insecticides.

The sex attractant of the female moth is a very complicated chemical mixture. The investigators noticed male moths flying to oak leaves, especially ones chewed on by oak leaf roller caterpillars. The males acted very excited, as if female moths were present. The idea occurred to the scientists that perhaps oak leaves contained some of the chemicals in the sex attractant mixture. When they analyzed the leaf substances, they found they were right. When they raised moths in the laboratory on wheat germ and alfalfa, however, the females could not attract males, and their bodies were proved to lack the attractant chemicals.

Other insects as well probably "borrow" sex attractants from plants. For example, two chemicals found in apple leaves are used in different combinations as sex attractants by more than a dozen apple tree pests. Biologists studying sex attractants in a number of insects have found that unless the larvae are raised on their natural food plant, the females, when they have developed, do not attract males. Studies such as these help scientists to understand how insects communicate with one another. By understanding the language of insect pests, they can develop ways of interfering with pest reproduction. Using such methods, we can eliminate harmful insects without poisoning helpful insects and other animals.

3

How Insects Pollinate Plants

One of the most fascinating examples of the give-and-take between plants and insects is the pollination of flowers. The sight of a bee crawling over a flower, busily collecting pollen and nectar, is familiar to almost everyone. The powdery yellow pollen contains the male reproductive cells of the plant. For a new generation of plants to be produced, the pollen must reach the female parts of a flower and fertilize it. The flower exudes nectar, and this acts as a lure for the insect, which visits it and carries some of its pollen away to another flower. Most animals have separate male and female individuals which must mate with each other to produce offspring. But most flowers contain both male and female parts. Why then does the flower "bother"—in many cases—to attract insects to carry its pollen away instead of using its own pollen to fertilize itself? The answer lies in the importance of maintaining variability within a species.

The most vital function of any species of plant or animal is reproduction. No matter how robust and

PINE

SORGHUM

BLUE GENTIAN

DANDELION

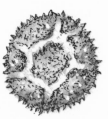

RAGWEED

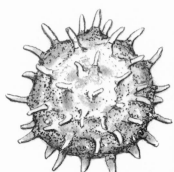

ROSE OF SHARON

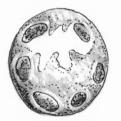

MORNING GLORY

SAGE

OAK

CORYDALIS

RUSSIAN THISTLE

healthy an individual organism is, it must leave healthy descendants or its kind will die out. Most organisms reproduce by mating with another individual of their own kind. This method constantly recombines the traits of the species, producing individuals which are not identical. Each individual possesses a unique combination of traits which make it most suited to live in slightly different conditions from every other. With each generation, environmental conditions may vary. It is important to have a variety of different individuals produced so that at least some of them can live and reproduce successfully, continuing the species. When inbreeding occurs—the mating of closely related individuals—there are far fewer variations. This sets up roadblocks against adjusting to changes in the surroundings.

Self-fertilization works for some organisms. Some plants, such as peas and beans, get along fine by relying mainly on self-pollination. But by far the majority of higher living things reproduce by "outbreeding" to other individuals of their kind. The elaborate variations in flower structure insure such cross-pollination. Some plants, such as grass, wheat, and corn, are pol-

Left: *Pollen under the microscope reveals a fantastic world of shapes and textures. By examining them a pollen expert can identify the plant that produced them, even if the grains are fossils millions of years old. Pollen produces the sperms that fertilize the female egg cells of angiosperms, or flowering plants.*

linated by the wind. They have small, uncomplicated flowers. A very few plants are water-pollinated. But the vast majority of plants rely on insects, birds, and sometimes bats to carry pollen from one flower to another.

Flower Structure

Although there are tremendous differences in the structure of flowers, a basic pattern exists. A simple flower has a circle of colorful petals. Outside the petals at the base of the flower is a circle of green leaflike sepals. The sepals surround and protect the flower before it opens. The sexual parts of the flower are located inside the petals. In the center is found the female organ, called the pistil. The enlarged lower portion of the pistil is called the ovary. The ovary contains the egg cells. The middle stalklike part of the pistil is the style, which is topped by a knob called the stigma. Arranged around the pistil are many male parts, stamens. Each stamen consists of a thin stalk called the filament, and a top piece, the anther. The anthers contain the pollen.

Flowers have many ways of avoiding self-pollination. Some have a tall pistil which sticks out above the stamens; an insect which is collecting pollen cannot brush against it. In other flowers, the pollen is ripe when the pistil is not, and any pollen which might be deposited on the stigma cannot pollinate it. Some plants

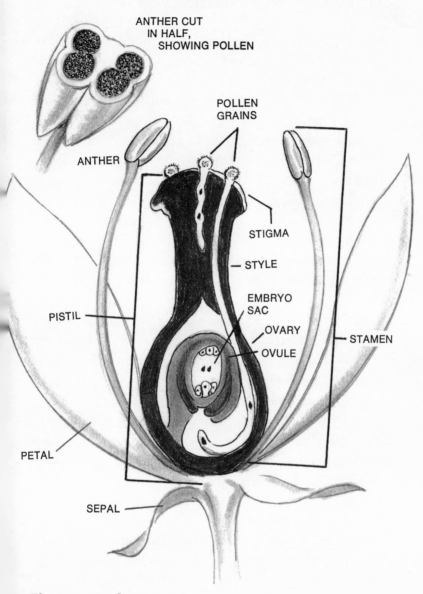

ANTHER CUT
IN HALF,
SHOWING POLLEN

POLLEN
GRAINS

ANTHER

STIGMA

STYLE

EMBRYO
SAC

PISTIL

OVARY

OVULE

STAMEN

PETAL

SEPAL

The parts of a flower

have separate male and female flowers. Others have separate male and female plants, making self-pollination impossible.

When a flower is pollinated, pollen is brushed off the pollinating insect onto the stigma. Many pollen grains are deposited at the same time. Each grain then grows a long tube down through the style into the ovary. Inside the ovary is a series of small ovules, each of which contains an egg. One male cell then combines with one egg cell to produce a "zygote." The zygote develops into an embryo. The other parts of the ovule change and grow, too, and the embryo plus the other parts contributed by the ovule combine to form the seed.

Although some pollinators, such as bees, eat pollen, most are attracted by the nectar the flower produces. Nectar may come from different parts of the flower. Sometimes it is present on the surface, as tiny droplets. Other flowers, such as the Christmas rose, provide nectar in little pitcher-like containers which radiate from the center part of the flower. Nectar is often hidden away in deep spurs, or hollow appendages, as in nasturtiums and columbines.

The location of the nectar is related to the kind of animal which pollinates the flowers. The scent, color, size, number, and complexity of flowers also depend on the nature of the pollinator. Most flowers specialize in attracting birds, bats, or a specific kind of insect. In this way, they can lure a pollinator which is efficient for a specific size and type of flower. In effect they discourage visits from inefficient ones which take nectar without pollinating in return.

Nectar Guides

If you look closely at a variety of flowers you will see that some, such as pansies, have little lines radiating from the center. Others have a ring of contrasting color around the inner part of the flower. These markings serve an important function. They point the way to the source of nectar, making it easier for flying pollinators to zero in on the right part of the flower as they approach. Some flowers have nectar guides which are not visible. They are stripes of odor rather than color, produced by special parts of the petal.

These nectar guides make efficient collection of pollen and nectar possible for insects. The insect uses up energy flying around looking for food. It is important to pollinators that flowers be well marked so that they waste as little time as possible searching. Flowers with distinct scents and/or bright colors will be easy to find, and nectar guides can help the insect focus its attention on the important part of the flower.

Once a pollinator has begun to collect from one kind of flower, it tends to stick to that type as long as it can find blossoms. This "flower constancy" increases the effeciency of collection for the pollinator. And without it, pollination would be much more difficult for flowers.

Because of flower constancy, the right kind of pollen is brought to a flower for fertilization. Some pollinators are more constant than others. Honeybees display almost complete flower constancy, while beetles may show very little.

Bees and Bee Flowers

Bees are the most frequent pollinators of flowers. While butterfly and moth larvae generally feed on plant leaves, most bees at all stages of their lives feed on pollen and nectar. There are many kinds of bees, and different flowers have evolved adaptations to pollination by different kinds of them. Bee flowers have certain characteristics in common, however. Since bees fly by day, bee flowers are open during the day. Bees have keen color vision, and their flowers are usually colorful. Bee color vision is different from ours, however; they cannot see red, but they can see ultraviolet colors which are invisible to us. Many bee flowers which appear white to us actually possess ultraviolet colors which we cannot see. In ultraviolet light, some white bee flowers show strikingly contrasted nectar guides as well, which are completely invisible to the human eye.

Some bee flowers provide only pollen. Instead of offering the bee sweet nectar, they produce an abundance of pollen which the bees collect and use in feeding their larvae. Wild roses, peonies, and poppies are examples of pollen flowers. Scotch broom is a pollen

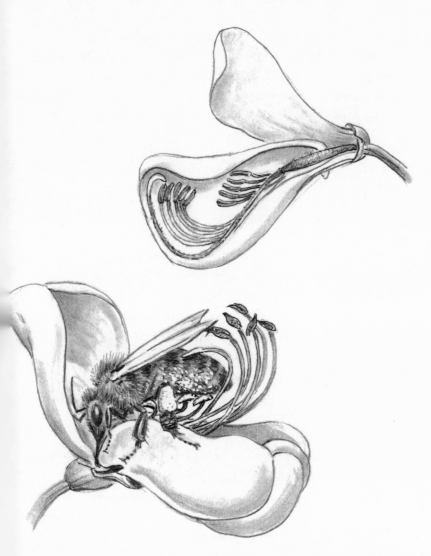

The flower of the Scotch broom in cross section, top, showing stamens. A bumblebee pops the flower open as it lands, getting brushed with pollen on both its belly and back.

flower which carefully guards its pollen. Until a bee visits it, the Scotch broom flower is closed. But when a heavy bee, such as a bumblebee, lands on the flower, the insect's weight springs the flower open. Five short stamens pop out and brush the bee's belly with pollen. For good measure, a bunch of long, curved stamens also reaches around over the bee and covers its back with pollen.

Other flowers have tricky ways of "making sure" their pollen is transported. The barberry flower has a tall pistil in the center. Nectar is secreted by glands near the petal bases. It collects around the base of the pistil. The anthers are curved outward against the petals. When a bee probes with its proboscis, or sucking organ, for nectar it will almost certainly touch the base of a stamen. Once touched, it curves inward and brushes the bee's proboscis with pollen. After a while the anther returns to its original position in wait for the next bee. The pollen on the proboscis will likely be deposited on the protruding pistil of the next flower the bee visits.

Some flowers are adapted to pollination by bumblebees. The bumblebee has a longer proboscis than the honeybee, so it can reach nectar deeper inside a flower than can a honeybee. Common red clover is a bumblebee flower. Although honeybees may visit clover, they are not effecient pollinators. When clover was introduced into New Zealand in the nineteenth century, no bumblebees were present. Until some British bumble-

The carpenter bee takes a short cut to nectar by punching a hole in the flower, with the result that the plant loses the opportunity to be pollinated.

bees were introduced, the clover in New Zealand could not produce a good seed crop.

Some flowers with deep nectaries must protect themselves from large bees, however. It is much easier for a bumblebee or carpenter bee to cut a hole in the base of the flower from the outside than it is to struggle inside the flower and lap up the nectar with their long tongues. When the bee uses this short-cut method, it does not pick up or transfer pollen. Robbed flowers have wasted their nectar. Some flowers are so easily robbed that scientists worry about their survival in certain areas. Others are very well protected. Thick flower bases make it hard for bees to cut through to the nectaries. Long, thin flower stems make it hard for bees to keep their balance on the flower. And, as we will see later, some flowers have the aid of ants in discouraging robbers.

Butterfly and Moth Flowers

Butterflies and moths are closely related insects which often collect nectar from flowers. Their long, tubelike proboscis can suck up nectar from down deep within the flowers. Since moths are night-fliers, moth flowers tend to open at night and be white or greenish in color. Some flowers attract the hawk-moths, strong fliers which hover in front of the flower while they gather nectar. Such flowers have long nectar tubes and no landing platform. Noctuids are moths which do land on flowers, and the flowers they pollinate have landing platforms and shorter nectar tubes.

Butterflies fly by day and are not very strong fliers. They need a landing platform, and most butterfly flowers grow in clumps rather than singly. This gives the butterfly plenty of landing space. Butterfly flowers are often white or purple and have pleasant fragrances. They provide plenty of nectar; it is deep within the flower where the butterfly can collect it with its long proboscis. Butterflies also visit the same flowers as bumblebees, especially in the spring. But some butterflies, such as swallowtails and whites, can see red. They pollinate flowers like wild pinks. Pinks hold nectar in central tubes which can be reached only by a long proboscis.

Most red flowers are pollinated by hummingbirds. Since birds are warm-blooded and larger than insects, they use up more energy flying from flower to flower. Therefore hummingbird flowers must provide them with a generous amount of nectar. They must also guard this treasure from other creatures. The red bird flower does not appear colored to bees and blends into

Some plants, such as honeysuckle, have no petals that are easy to land on, but this doesn't deter the hummingbird hawkmoth, which is an expert at hovering in the air at one spot.

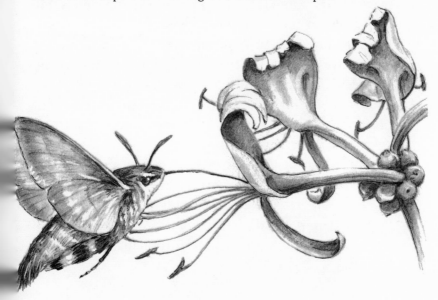

the background vegetation. It also is usually odorless and has the nectar hidden deep within where only the long tongue of a hummingbird can reach it.

Pollination by Ants

Scientists have argued through the years whether ants can ever be efficient pollinators of flowers. Some have said that ants are too slow, being walkers and not fliers, and that ants wander from one kind of flower to another stealing nectar and are never constant enough to pollinate properly. Ants have smooth bodies with only a few hairs to which pollen can stick. But a definite case of ant pollination has been described.

The plant involved is a small, low-growing one found in dry areas of the western Cascade Mountains. It has very small, funnel-shaped flowers with no detectable odor. Nectar is produced by five little glands at the base of the funnel. The silky ant actively collects nectar from the flowers. As the ant reaches down into the flower, sticky pollen is deposited on its head. Like many other flowers, these blossoms produce pollen before the stigma is ready, thus avoiding self-fertilization. As the ant travels from flower to flower it encounters female flowers with receptive stigmas. The pollen on its head then fertilizes them.

Since the ant usually visits only one flower of each plant, cross-pollination is very likely. The ants are constant as well, collecting nectar from many flowers on one trip. The flowers are too small and yield too little nectar to attract flying pollinators, yet 85 to 100 per cent of the flowers produce seeds. This proves conclusively that ants are capable of being efficient pollinators. Probably other low-growing plants with small flowers which live in dry areas are pollinated by ants as well.

Other Insect Pollinators

Flies, gnats, and beetles can also pollinate. Some flies, such as hover flies, mimic bees so well that one must look closely to see what they really are. Instead of buzzing constantly from place to place, these hover flies stay in one spot in the air and then dart away. Beetles pollinate some flowers such as magnolias. They are attracted by the strong scent and may chew on the flowers as well as collect the accessible nectar.

The more interesting beetle and fly flowers, however, attract their pollinators by "deception." The spotted arum lily—sometimes called the cuckoopint—is a fascinating example. The "flower" of the arum lily is not a single flower, but rather a collection of male and female flowers associated with other parts. A large, modified leaf is wrapped around the central stalk of the arum "flower." The lower part of the stalk is covered with small female flowers which secrete a sweet, slimy liquid. Above the female flowers is a circlet of bristles, with the little male flowers above them. A bit higher on the stalk is a region of long, hairlike bristles. The stalk then continues out of the surrounding leaf.

When the arum flowers, the leaf unfolds and the top third of the stalk, which is called the appendix, heats up as much as 25° to 36° above the temperature of the surrounding air. This heat helps intensify the strong rotting-meat aroma of the appendix. We may find the smell repulsive, but many flies, gnats, and beetles love it. They fly to the arum, but the long upper bristles keep out the bigger beetles and flies. The smallest beetles and gnats can get through, but once they do they are in for a surprise. The inner walls are covered with slippery oil droplets which make it impossible for them to hang on. They slip and fall down inside the flower. They cannot climb out. If they crawl up over the female flowers, their way is blocked by the lower row of bristles.

Life is not so bad for them, however, since they like to eat the sweet liquid on the female flowers. If they have previously been "guests" of an arum, they will leave pollen on the flowers and pollinate them. All night the arum remains closed, but as night passes, the male flowers ripen and drop their pollen in a constant, slow rain upon the trapped insects below. By morning the bristles have lost their stiffness and the little prisoners can crawl up the stalk to freedom. None the worse, but covered with pollen, they fly away. If another arum should attract them with its rotten aroma, they will again become prisoners and pollinate its female flowers.

There are many plants of different families which have similar techniques that attract and fool pollinating beetles and flies. The flowers of the Dutchman's pipe lure flies, while some arums attract dung beetles.

Depending on Each Other

Two cases are known in which a plant and insect depend completely on one another for reproduction. The yucca moth needs the yucca plant just as much as the plant needs the moth. The moth pollinates the plant, while the plant nourishes the moth caterpillars. Neither species can survive without the other.

The yucca is a plant of the southwestern American deserts. The leaves form a large, spiny mound out of which a tall flower stalk grows each spring. Each stalk has many large white flowers which hang down like bells. When the flowers are ready to be pollinated, they open at night, releasing a powerful aroma. The mated female moth is attracted. She goes to the stamen of a flower and scrapes off the pollen with special "palpi" under her chin. She kneads and shapes the pollen until she has a smooth ball bigger than her head. She holds tightly to it with the palpi. Off she flies, carrying the pollen ball, to another flower. She goes to the ovary of the second flower and lays her eggs inside among the ovules. Then she climbs to the end of the pistil and jams the pollen ball into a hole at the end of the stigma.

With this act she pollinates the yucca flower, insur-

The yucca plant and the yucca moth depend on each other completely for survival. Above, she collects pollen from a stamen; below, she pushes a ball of this pollen into a hole at the end of the stigma, the top of the female organ.

ing food in the form of yucca seeds for her offspring. Each caterpillar will eat about twenty seeds as it grows, and the yucca flower produces around two hundred seeds. Since the moth lays four or five eggs, plenty of seeds are left to make new yuccas. The female moth cannot know what she is doing when she pollinates the yucca flower. Yet she always does so, pollinating several flowers during her one evening's work, ensuring still another generation of moths and yuccas. Shortly after completing her task, the moth dies.

The Fantastic Figs

Still more amazing in its complexity and timing is the relationship between fig trees and fig wasps. With few exceptions, each fig species has its own distinct species of wasp pollinator. The shapes and sizes of the fig and wasp parts in each case have evolved together so that only one wasp "fits" one fig species. This system works well, however, for there are more than 900 species of figs. These plants are found all over the world in regions where frost and deserts do not occur. Most are trees, but some figs are shrubs or vines. Not all types are edible.

The fig is a peculiar sort of fruit. Instead of developing from a fertilized flower, as do most fruits, the fig is actually a sac *inside* of which the flowers develop. First the female flowers mature. Several weeks or even months later the male flowers develop. This makes self-pollination impossible. All the figs on a single tree are in the same stage at the same time, while nearby trees may be in different stages. Year-round reproduction is essential so that some male-stage fruits are available at the same time as female-stage fruits on other trees.

The fig contains two types of female flowers. One kind has a long style, while the other kind has a short one. The female wasp pollinates both kinds of flowers. She also lays one egg in each of the short-styled "gall flowers." The larvae feed on the tissues of the gall and pupate there. They reach maturity just when the male flowers are ripening and the fig is beginning to soften. Fewer than half of the wasps are males. They look so different from females that once they were thought to be completely separate species. The male is wingless, has reduced eyes and antennae, and pointed, well developed jaws. The males emerge from their galls first. They crawl about in the dark, cramped cavity of the fig, finding females and fertilizing them. After all the females are fertilized, the males cut a tunnel out of the fig.

Right: *Edible figs depend on a complicated pollination process. 1, a female wasp enters a caprifig; 2, it lays its eggs in the ovary of the interior flower; 3 and 4, males and females develop from the eggs and a male fertilizes a female; she escapes from the caprifig and (5) flies to an edible fig, entering it and pollinating its flowers, which are exclusively female.*

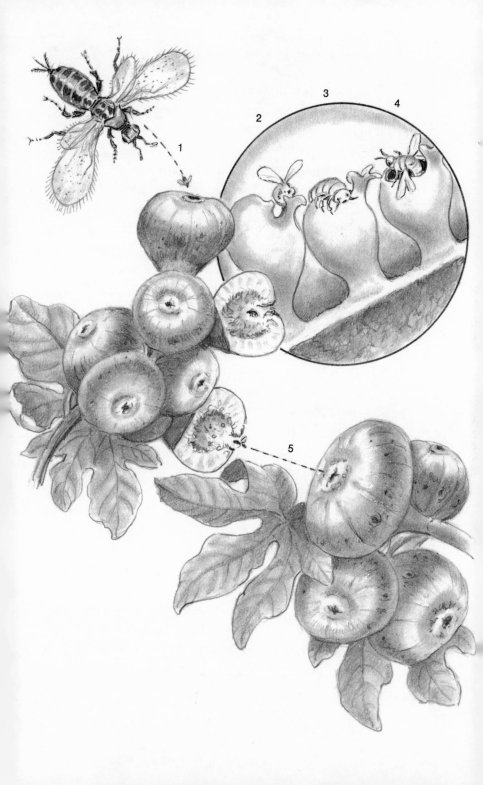

Scientists recently made an interesting discovery about one kind of cultivated fig. They found that when the male wasps are active, the concentration of carbon dioxide gas, produced by respiration of the fig, is high. When the males cut the tunnel to the outside, air flows in and the carbon dioxide concentration decreases. Then the females become active. Males can function only with a high concentration of carbon dioxide, while females are active only at normal carbon dioxide level.

Once females of this species come out of the galls, they go to the ripe male flowers. With their jaws they crumble the pollen and then transfer it to special pollen pockets on their bodies. The pockets pick up the pollen very quickly. The entrance to the pocket is sensitive and opens a crack whenever something touches next to it. Then it closes up again. When her pouches are full, the female wasp leaves through the exit made by the males. Unlike the male, she has well developed antennae and eyes and strong wings. She flies off in search of a fig tree at the proper stage of development. Scent probably guides her.

Upon finding a fig, she locates the tiny entrance channel and struggles inside. By the time she reaches the cavity, her wings and sometimes even part of her antennae have been torn off. But she won't need them any more.

The female wasp now lays her eggs. At the end of her body is a long egg-laying tube called the ovipositor. She jabs it down through the stigma into the style of each flower. If the flower is a short-styled gall flower, her ovipositor can reach the ovary and she lays an egg. If it is a long-styled flower, her ovipositor is too short. Since she cannot reach the ovary she does not lay an egg. After she is finished laying eggs, the female wasp rapidly unloads the pollen with her forelegs and strokes it across the tops of the stigmas, pollinating them. Once that is done, she has fulfilled her destiny. Within a few days she dies inside the fig.

Many new facts have recently been learned about figs and their wasps. The only difference between gall flowers and seed flowers is the length of the style. If an egg is not laid in a gall flower but it is pollinated, it develops a normal seed. The long-styled flowers can support the growth of a wasp embryo just as well as a gall flower can. Scientists used to think that pollen clinging to the wasp's body or eaten by the wasp was used to pollinate. But now the special pollen pockets have been found on the bodies of many fig wasp species. Other kinds have pollen baskets on their legs.

The pollinator of the common edible fig, however, does seem to carry the pollen in her digestive tract. Pollination in the edible fig is more complicated than in most other kinds, because there are two types of fig trees. One bears figs with short-styled gall flowers which are followed in the normal fashion by male flowers. It is called the caprifig. The other type has only long-styled female flowers; it never develops male flowers. These become the edible figs.

Figs have been an important food in the Mediterranean region since prehistoric times. The connection
between wasps and figs was first noticed by the great
ancient Greek philosopher Aristotle. Edible fruit is
produced only when caprifigs are growing near the
edible figs. So a few caprifig trees must be planted between the female trees; or branches of caprifigs, cut at
the right time, are hung in the orchards.

When the female wasp emerges from a caprifig, she
is equally attracted to the female edible figs and to the
caprifigs. Chance determines which type she enters. If
she enters a caprifig, she can easily lay her eggs through
the wide, funnel-shaped stigma. But if she enters an
edible fig, all her egg-laying attempts are frustrated by
long, narrow stigmas. In either case she pollinates the
flowers. There are some varieties of the edible fig, such
as the Smyrna fig, which will develop into soft, luscious
fruit without being pollinated. They can even grow
in England, where the climate is very different from
that of normal fig-growing regions and no fig wasps are
found.

Extra Warmth

Scientists are only beginning to study another way
in which some plants and insects benefit one another.
Life in the high arctic regions of the earth is very chal-
lenging. With a growing season of only about six weeks,
plants need all the warmth they can get. Certain arctic

flowers of the rose and poppy families are bowl-shaped. Their petals reflect the sun's heat down onto the female parts of the flower, much as a parabolic reflector concentrates the sun's heat on the center part of a solar stove. This speeds up seed development. Some such flowers even turn slowly, following the sun so that they gather every precious bit of heat possible.

Arctic insects must also grow, mature, and reproduce during the short summer. They, too, benefit from extra warmth, which hurries along their growth. They do this by taking advantage of the heat collected by the arctic flowers. Instead of busily flying from flower to flower, such insects take their time and bask in the warm flower centers. Not only do they gain heat in this way but they can also obtain food in the form of pollen and nectar. The flowers gain, too, for their greater warmth increases their attractiveness to potential pollinating insects. Perhaps future research may show that the "parabolic reflector" mechanism exists even in some plants of temperate zones.

4
The "Ingenious" Orchids

Everyone has heard of the exotic orchid. It is looked upon as a special flower because of its great beauty and tropical habitat. The orchid is special to the biologist as well. Its life history is just as exotic as its appearance. There are unique and unusual aspects of orchid life at all stages, but here we are concerned only with the wonderful variations in orchid pollination.

The orchid family is the largest of all plant families. About 10 per cent of all flower species are orchids. Contrary to what we imagine, orchids are not exclusively tropical. They are found from the Arctic region south through the temperate regions and tropics all the way to the edge of the Antarctic.

They survive in a variety of habitats, including deserts and alpine meadows as well as tropical jungles. The size of orchid flowers varies considerably. The smallest is less than one-sixteenth of an inch long, while the largest grows to over a foot in size. Their colors span the rainbow and more, and their shapes are of endless variety. Although a tremendous range of pollinators has been claimed for orchids, including frogs and snails, only insects and hummingbirds have so far been confirmed by scientists as orchid pollinators.

The amazing variation in color, size, and shape of orchids indicates the wide variety of pollination mechanisms found in this family. Orchids are most frequently pollinated by bees. Sometimes many bee species will visit one kind of orchid. In other cases, only one kind of insect is attracted to a certain species. Taken all together, bee-pollinated orchids provide the insects with a great variety of ways to get the bee to the right part of the flower at the right time.

Scientists estimate that from one-third to over half of orchid species do not provide food to their flying visitors. Even those which reward them with nourishment sometimes do so in strange ways. Orchid pollen is uncollectable, as you will see later. Many orchids do provide nectar. But others have false pollen in the form of modified hairs which bees may collect. A few orchids provide bees with wax that they use for nest-building. But many orchids "trick" pollinators and offer no food. Instead, the flower mimics a food plant, other insect prey, or even a mate. It may have false stamens or nectaries which provide no nourishment. Some attract flies with rotten odors.

The Orchid Flower

Orchids belong to a different group of plants than do most familiar garden flowers. They are more closely

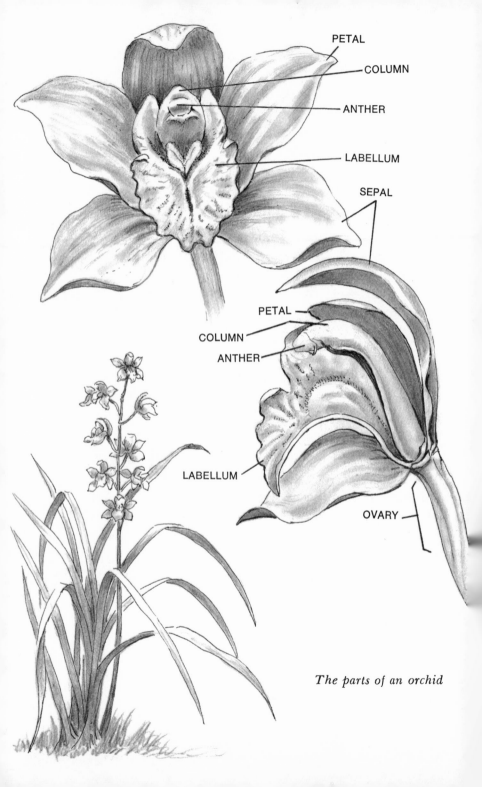

PETAL

COLUMN

ANTHER

LABELLUM

SEPAL

PETAL

COLUMN

ANTHER

LABELLUM

OVARY

The parts of an orchid

related to lilies and grasses than to daisies and geraniums. The structure of the orchid flower is therefore different in several ways. The sepals of orchids are usually colorful and resemble petals. There are three sepals and three petals in each flower. One petal, which is usually at the bottom of the flower, is different from the other two. It may have fancy ruffled edges and complex patterns in bright colors. It often serves as a landing plaform for pollinators and is called the lip, or labellum.

The biggest difference between orchids and other flowers is in the reproductive parts. Instead of being a swelling at the base of the style as in many flowering plants, the ovary is hidden away in the stem of the flower. The stamen, stigma, and style are united in most orchids to form the "column." The column usually arches up above the lip. The pollen is different, too. Instead of separate pollen grains, the orchid has pollen packets called pollinia. One pollinium may contain over 100,000 pollen grains.

Pollination in orchids is a very precise process. When a pollinator visits the flower, characteristics such as scent, location of nectar, or structural "tricks" of the flower cause some characteristic part of the pollinator's body to come in contact with the column. The pollinia are glued on and carried to another flower. If the pollinator already carries pollinia from another flower, they are deposited on the stigma. The stigma of orchids is reduced to a shallow depression of the inner side of the column. It secrets a sticky, sweet liquid in which the pollinia are deposited. The pollen grains germinate in this liquid and grow into the column to the ovary.

There are several advantages to this method of pollination. Since the pollen is packed into pollinia, none is wasted by being scattered about or eaten. One visit by a pollinator can fertilize thousands of ovules. Even if a bee visits several different kinds of orchids it can still accurately pollinate each kind, since the pollinia will be located on different parts of its body. Only a pollinium in the appropriate location can be picked up by the stigma of a particular species of orchid.

Simple Pollination

First let's consider how the pollen is transferred in a relatively ordinary kind of orchid. Cattleya is an orchid which provides bees with nectar. The nectar tube is deep inside the flower, however, and the bee must crawl in between the labellum and the column to get it. As the bee backs out of the flower, its back rubs against the sticky secretion of the stigma. Some of it sticks to the bee. As the bee moves out farther its back brushes against the pollinia, which are then cemented to it by the secretion. The bee carries the attached pollinia on to another Cattleya flower. When the bee is leaving that flower, the pollinia are deposited in the cavity of the stigma. As the bee continues out of the flower, it is again wiped with sticky material which picks up the pollinia of that flower. Each time the bee visits another Cattleya it pollinates it and picks up new pollinia.

False Food

One orchid attracts both bees and flies expecting a meal. After landing on the labellum, an inquisitive insect walks down the wide slide formed by the sides of the lip. It heads for a patch of green spots which apparently imitate food. But the region near the spots is slippery, and the insect slides into the deep pouch of the labellum. The only way out is to crawl up the other side and through a small opening under the stigma, then out one side of the column. As the insect squeezes by the anther, a pollinium is attached to its back. When the insect visits another flower, it deposits the pollinium as it crawls under the stigma.

Another orchid has protected false nectaries. Bumblebees struggling to get to what they think is nectar merely end up with pollinia attached to their heads. A stalk attached to the pollinium bends over in front of the bee's head, placing it in a perfect position to be attached to the sticky stigma of the next flower.

The grass pink of the eastern United States and Canada plays a trapdoor trick on visiting bees. The labellum is thickly covered with yellow hairs looking for all the world like a mass of pollen-covered stamens. Upon landing, however, the bee is thrown onto the column by the collapse of the hinged labellum.

Mimicking Insects

There are two ways an orchid "fools" a bee into thinking the flower is another insect. Some male bumblebees set up territories. Any entering insect is attacked. Certain orchids which to our eyes may only vaguely resemble insects have arched stems. Even a mild breeze makes the flowers dance. The resident male bee is apparently fooled into thinking an insect has invaded and attacks. He flies straight at the flower and bangs it with his head. In return he gets a pollinium right between the eyes. Whenever the breeze moves the flowers, the bee attacks, pollinating flower after flower.

Several orchids use the other technique. The flower "pretends" to be a female insect. Its odor and appearance attract male bees and wasps which try to mate with it. Ophrys orchids belong to this group. They are found mainly around the Mediterranean Sea, but a few also occur in England and Sweden. The Ophrys flower is modified in several ways for its role as a pretended female insect. The labellum is thicker and stronger than in most orchids. It has strange color patterns in dark brown, purple, white, and metallic blue. Shiny false nectaries resemble the eyes of a bee. Projections

Left: *In the grass pink the labellum, which is hinged, falls over on the bee as it lands, pressing it against the reproductive parts.*

or bands of color may mimic antennae or wings. Some Ophrys flowers look startlingly like insects. With other species, the resemblance is slight to human eyes.

But the vital stimulus to the male bee is the flower scent. Female wasps and bees produce sex attractants. Apparently the Ophrys flower imitates these, even improves upon them. For once a male bee has landed, he is reluctant to leave, even though his mating attempts are unsuccessful. He presses his body down on the labellum. His face pushes against the column in the process, and pollinia are attached. When he finally leaves, he takes the pollinia with him to the next Ophrys flower which beckons.

Ophrys has adjusted its flowering period to the bee's life cycle. The pollinators are solitary bees and wasps. The males emerge a few weeks before the females. Ophrys blooms mainly during this time while the males are flying and the females are not. Thus they do not need to compete with the real thing.

The Remarkable Golden Bees

In the American tropics live the golden bees. Some are golden in color, but others are black, metallic blue, or green. They belong to a group called the euglossines. These beautiful bees are solitary. The female cares for her own young, although sometimes several females will nest together. The males live a carefree life, often flying long distances. They may survive as long as six months, which is a long life for a male bee.

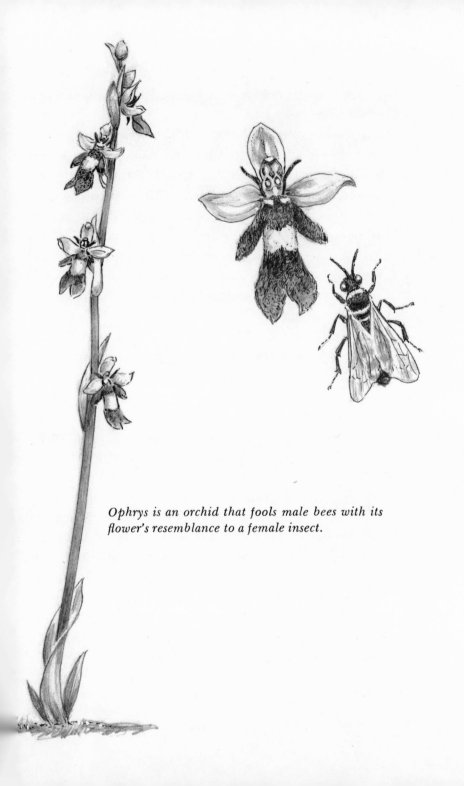

Ophrys is an orchid that fools male bees with its flower's resemblance to a female insect.

The strange behavior of male euglossines on orchid flowers has been known for a long time. Only recently has a reasonable explanation for it been found. Many orchids of different species are pollinated by male euglossines. The bees are attracted by the scent of the flower. When a male bee visits such an orchid, he brushes the surface with special bristle pads on his front legs. After a while he flies up above the flower. While hovering there, he quickly brushes his legs together. He lands again and repeats this strange performance. He may remain for ninety minutes at one flower, landing and hovering, landing and hovering. We now know that when he brushes, he is collecting odorous substances which ooze out when the flower is scratched. While hovering, he is transferring them to complicated enlargements on his hind legs where they are stored.

Why should he do this? The chemicals do not provide food. They do not attract females. They do not contain any vital trace nutrients like vitamins. Scientists puzzled over the function of these chemicals for years. The scents must be important, since the complex enlargements of the hind legs have no other use. The brushes are used only for collecting them. And a great variety of orchids depend on the needs of euglossines to insure their pollination.

Finally C. H. Dodson, who has spent a great deal of time studying euglossines and orchids, came up with a reasonable explanation. Although not completely proven, at least his theory makes sense. At mating time, male euglossines pick out perches and put on fancy aerobatic displays to attract females. Some species have diving and landing routines, while others fly in zigzag patterns. They buzz during the display, which apparently attracts the females. The male that establishes a display area marks it with the orchid fragrances he has gathered. Other male bees are attracted and display with him. The display of half a dozen bees should be more noticeable to a female than the display of a single bee. By attracting other males to his perch, the first male bee increases his chances of attracting mates.

Male euglossines are certainly put through the mill by some orchid flowers. But they keep coming back. Stanhopea orchids take advantage of the "drugged" condition of the bees. While visiting orchids to collect their scents, these bees become less wary and their movements may be slow and clumsy. Whether this is due to effects of the orchid scent or merely to single-minded concentration by the bee is not known. But this lack of coordination is the key to pollination of many euglossine orchids. The Stanhopea orchids make the bees fall, slide, or get dunked while pollinating, and the bees never seem to learn a lesson from their experiences.

The flowers of some species face downwards. The sepals and petals are curved up, and the labellum and column are close together, pointing down. The male

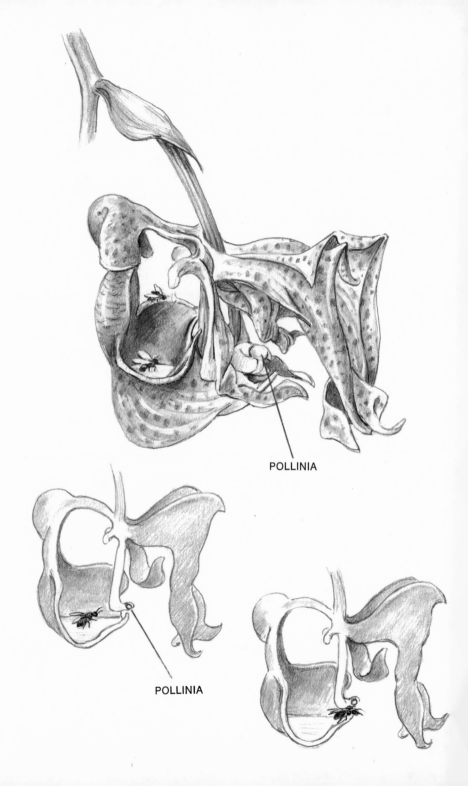

POLLINIA

POLLINIA

bee enters the flower near the base of the lip, where the space between lip and column is wider than at their tips. He scratches at the lip. When he tries to fly away, the column gets in his way and he falls through the narrow chute formed by the lips and column. As he falls, the projecting pollinium is attached to his body.

One of the strangest euglossine flowers is Coryanthes. The petals and sepals are folded up out of the way in the mature flower. They resemble somewhat the sails of a boat. The labellum is very complicated in structure. The base is hooded, and the lip continues as a chute down to a deep bucket at the bottom. The bucket is filled with a watery liquid secreted by two glands above it. The fluid drips into the bucket, like water from a leaky faucet. Bees of several species are attracted by the powerful scent produced in the hooded upper part of the flower. While scratching, somehow they fall or are knocked off and slide down the chute into the fluid-filled bucket. The only way out is over the edge of the lip. The column hangs down in front of the lip, forming a tunnel through which the soggy bee must crawl to escape. The pollinia are deposited on him as he crawls out of the trap. Apparently undeterred by his experience, the bee flies on to another Coryanthes and repeats the whole sequence, this time pollinating the flower.

Left: *The "bucket orchid," Coryanthes, forces its pollen onto a bee as it struggles out of an unexpected bath.*

Many orchids use different techniques that enlist eu-glossine bees in their pollination. Because the place-ment of pollinia is so precise, one bee can pollinate different species during one fragrance-gathering flight. He does not have to be constant to one kind of flower. Since these bees sometimes fly many miles, they may be vital to the survival of some widely scattered orchid species. Once the pollinium is attached, it does not matter if the bee flies a long way and visits other flowers as he goes. As long as he eventually reaches another blossom of the same species, the pollinium will be there, attached in just the right way for correct pollination.

Other Orchid Pollinators

Butterflies, moths, and birds pollinate many orchid species, but beetle pollination is rare. The same kinds of flower types for different pollinators exist in orchids as is other flowers. Bird-pollinated orchids are often red, butterfly orchids have a landing platform, and so forth.

Quite a few orchids are fly-pollinated as well. They usually have rotten aromas and trap flies in cup-shaped lips, providing only one way out past the column. Some fly-pollinated orchids have movable parts which throw the fly either down into the lip or against the column.

Variations in orchid pollination seem endless, and new ones may be discovered tomorrow. Some scientists suspect other kinds of orchid "deception." For example, some orchid flowers bear a striking resemblance to caterpillars. It may be that they fool parasitic wasps into landing. Such wasps lay their eggs in caterpillar bodies. If the orchid imitates a caterpillar, the wasp could pollinate it in her attempts to lay eggs.

5

The Strange Relationships
of Ants and Plants

Ants are among the most common and familiar insects
in the world. They can be found living almost every-
where, except in the most harsh environments. Ants
live in colonies which may contain a few individuals or
millions, depending on the kind of ant. Each colony
has a queen ant which lays all the eggs. The other ants
are also females, but they cannot lay eggs which hatch.
Most of the ants are workers; they collect food and care
for the queen and the young. Many ants also have
soldiers in their colonies which have very large jaws
and defend the nest against enemies.

Although most ants live in the ground, digging
chambers and tunnels to house the colony, other ants
nest in the trees or inside plants. Some, such as army
ants, do not build a nest at all, but live always on the
surface of the ground. At a favorable time of year, ant
colonies produce king and queen ants with wings.

These leave the nest and form swarms with the kings and queens of other colonies. They mate and the males die. The young queen loses her wings and starts a new colony. She takes care of the first eggs and larvae, but after some workers have been produced she retires to the sole task of egg-laying, and the workers take over care of the colony.

Most ants eat insects or bits of dead creatures. But many also feed on plants to some degree, and a few ants consume plants exclusively. A common source of food for ants is the nectar secreted by plants outside of flowers. Such extrafloral nectaries are common in plants. Although scientists used to feel that they served merely as sites for release of waste materials, many now feel that these nectaries are useful to the plant in other ways. By attracting ants, they encourage the insects to explore their branches and perhaps attack and remove plant-eating insects. The ants may even serve as scarecrows, discouraging some plant-eaters from attacking the plant at all.

The trumpet creeper is a vine with large clusters of bright orange flowers, commonly found in the eastern and midwestern United States. In addition to the abundant nectar of the flowers, which attracts hummingbirds and bumblebee pollinators, the trumpet creeper makes another kind of nectar which attracts ants. The glands producing it are very small, like little dots on the leaf stems, flowers, and developing fruit. Different kinds of ants gather this nectar. Some kinds

The trumpet creeper attracts three kinds of pollinators with two different types of nectar.

are better protectors of the plants than others. Vines visited by one quite aggressive ant species seem free of insect damage. But when a more timid kind of ant is the chief nectar gatherer, some of the flowers are robbed by bumblebees. Even these ants, however, probably provide some protection.

Some seeds are just right to be picked up and carried away from the parent plant by ants. Such seeds have an especially nutritious coating which the ants eat, leaving the seed unharmed. In this way they are dispersed, and the new plants will grow uncrowded, away from the parent plant.

Grain-Gatherers

The harvesting ants have been subjects of great human interest since ancient times. Although these ants do eat other kinds of food when it is available, they collect and store plant seeds for use when other food is scarce. They are found in dry areas in many regions of the world. Dryness is probably necessary for their success, for seeds will sprout if there is too much moisture in the air. Some harvesting ants specialize in certain kinds of seeds, but most collect whatever ripe ones they can find. These ants live in underground nests and store the seeds in special chambers. If the seeds get wet during a rainstorm, the ants bring them to the surface and dry them out before taking them into the nest again. If it is so wet that the seeds do

Harvester ants collect seeds and place them in underground storage chambers.

sprout, they are thrown away into a garbage dump not far from the nest entrance, which already contains seed husks removed earlier. Some harvesting ants are said to bite the seeds in such a way that they cannot sprout. But often, healthy seeds end up in the garbage dump, and a ring of plants, often quite different from the surrounding ones, sprouts around the nest. Most harvesting ants remove all plants for a few inches around the nest hole. This probably allows the sun to beat down and keep the nest as dry as possible. Many kinds also build up a crater in which they hollow out storage areas for the grain and chambers for raising their young.

Some scientists think that harvesting ants are important to plants, for they carry the seeds to new locations where many of them sprout. But others disagree, saying that the harvesters' influence on plant distribution is too small to be of any importance. In any case, the plants are certainly important to these ants, for they enable them to live in hot, dry areas where few other insects can survive.

Ants as Plant Enemies

Although the activities of most ants in eating insects makes them friends of plants and man, some ants definitely do serious damage to plants. One of the most interesting relationships between insects causes great harm to plants. Some insects, such as aphids (more will be said of them later), leaf hoppers, and scale insects, live by sucking the juices of plants. Since the amount of juice is very great and is probably taken up by the insects with little effort, much of it passes through the animal's body without being used. These insects thus produce a sweet liquid waste called honeydew. Ants of many kinds collect honeydew from aphids or from the surrounding leaves where it has fallen.

Groups of aphids with ants scurring around them on the undersides of leaves are a common sight in gardens. When an ant is collecting honeydew from the aphids, it strokes the aphid's abdomen with its antennae, and the aphid gently pushes out a drop of honeydew. The ant sucks it up and goes on to the next aphid. If that aphid has just given up a drop, the ant will have to go on to still another and return later.

In some cases, the ant and the aphid live in very close harmony. The corn-root aphid is well cared for by its ant associates. The ants keep the aphid eggs during the winter. When they hatch, the aphid nymphs are carried to the roots of nearby food plants by the ants. If the plants are uprooted, the ants move the aphids to new roots. If the nest is attacked, the aphids and their eggs are protected by the ants as integrated parts of the ant colony. They are cared for just as if they were part of the ant brood.

Other species of ant do not just collect honeydew from aphids they happen to run across. They actually take care of certain "herds" of aphids, keeping them

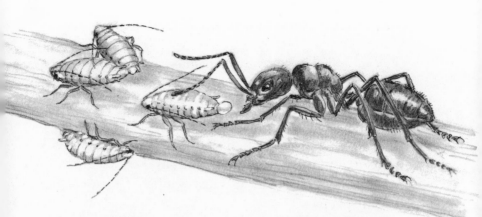

Ants and aphids have some unusual relationships. Above, ants eat a liquid called honeydew from aphids; below, the corn-root aphid is cared for by ants as part of the colony.

clean and attacking their enemies. In other cases, the relationship goes even farther, and the ants build shelters of silk or other material over their aphid herds. Some ants may carry the aphids from place to place. The better care given to the aphids, the worse pests they are to humans. When attended by ants, aphids can multiply at an alarming rate, sapping the strength of many plants.

Insect Gardeners

One very abundant family of ants owes its success to gardening skill. The most interesting and economically important ant gardeners are the leaf-cutting ants. These pests live in colonies of as many as two or three million individuals. Their leaf-cutting activity can destroy weeks of a human gardener's hard work overnight. Many residents of the American tropics have been horrified to go out in the morning and see the last shreds of their gardens disappear underground in the jaws of leaf-cutting ants. Plantation owners hate these ants, too, for they can cause serious damage to coffee and citrus trees.

Leaf-cutters do their damage by cutting out bits of leaves and carrying them down into their underground nests. For hundreds of years, people wondered what the ants did with all those leaves. Most felt the ants ate them, but others imagined they were used as roofing material for the insects' nests. A hundred years ago the truth was discovered by a scientist named Belt, who studied many interesting ants. He learned that the ants use the leaves to feed gardens of fungus, which they tend with great care. Since Belt's time, other scientists have studied the leaf-cutters, and now we know many details of their behavior.

The colonies of the more spectacular leaf-cutters contain a queen, soldiers, and workers of different sizes. The smallest workers are indeed minuscule, for the queen is more than six hundred times their weight. More than half the colony consists of these small workers which tend the queen, the ant brood, and the fungus garden. The middle-sized workers go out and cut leaves. They are about four times as big as the little workers. There are also a few large workers and some soldiers in the colony, and these are much bigger than the other workers. The queen is the largest ant in the colony, about four times as big as a soldier.

When the leaf-collectors go out to cut leaves, often some of the small workers accompany them and ride back on the bits of leaves, which the cutters hold high above their heads like umbrellas. They are not merely hitch-hiking, but are hard at work licking the bits of leaves to clean them and prepare them for use in the fungus garden. When the workers get back to the nest, they deposit the leaf pieces and go out for more. Meanwhile, workers in the garden cut the leaves into bits and chew their edges until they are mushy. They then "manure" the leaf fragments with drops of clear anal

Leaf-cutting ants carry off leaf chunks, above, and after storing them in underground gardens, harvest the fungi that grow on them.

fluid and insert them into the spongy mass of the fungus. New pieces are added around the garden edges. After being tucked in, tufts of the fungus are carefully planted on the new leaf bits by the tiny workers. The fungus grows quickly, for by the next day the fungus covers most of the leaf piece.

The fungus is the only food of the leaf-cutter ants, and the fungus is never found living outside the ant colony in nature. Thus the two are completely dependent on one another. The growing tips of the fungus produce rounded swellings which the ants pluck off and eat or feed to their brood. Only one special kind of fungus grows in the garden, despite the fact that the ants go in and out of the nest constantly bringing in spores of other fungi and bacteria. If a nest is abandoned for some reason, other kinds of fungi and bacteria very quickly take over the garden, overwhelming the ant-fungus.

How do the ants keep unwanted organisms from growing in their gardens? When scientists grow a single fungus species in the laboratory, they must be very careful to avoid contamination, but for the ants it seems to be easy. The ants' secret, however, is the same as the scientist's. They are extremely careful with the leaf fragments and with their own bodies. The leaf bits are always thoroughly licked clean, and the ants frequently groom their own bodies and those of one another. Before starting a new garden area, they carefully clean out the chamber, licking roots and stones clean. The new garden is started on a carefully cleaned surface, preferably of a rock or a root. The area around the fungus is kept clear of other material, lessening the chances of contamination. If a strange kind of fungus should take a foothold despite the care of the ants, it is plucked out and removed quickly by them. They seem to have developed the aseptic technique of the modern operating room long before there were any hospitals.

Colonies of leaf-cutters produce new queens and males before the rainy season starts. After a rain heavy enough to wet the soil, the queens and males leave the nest and fly off, often forming large swarms. After mating, the female digs into the soil, producing a small cavity. She removes a fragment of the precious fungus from a special pouch behind the mouth and carefully tends it. As the garden grows, she lays her first eggs in it, and a new colony is on its way.

There are also species of beetles and termites which tend fungus gardens, but none of them actually plant the fungus as do the leaf-cutting ants. Relatives of leaf-cutters tend fungus as well, but grow it on bits of rotting plants or insect droppings instead of leaf pieces. Their colonies are simpler in organization and have only one size of worker.

Ant Plants

In warm parts of the world, plants are found which are always or almost always inhabited by ants. The plants seem to be "made to order" as ant homes, with hollow stems or thorns which make ideal nesting places. Clearly, the ants gain something from this arrangement, but does it do anything for the plants?

Ever since the nineteenth century when Belt and other scientists discovered these strange associations between ants and certain plants, arguments have flown back and forth about their meaning. Some experts felt that the ants were taking advantage of the plants and that the plants got nothing out of it. But others were convinced that the plants benefited as well. They pointed out that in most cases the ants were especially vicious and ready to attack any intruder; that they were very effective in chasing away other insects or browsing mammals which might try to feed on the plant. These disagreements went on for many years before someone finally decided to study one of these relationships to settle the question.

The Ant Acacias

Acacias are plants found in Australia, Africa, and North and South America. About 10 per cent of the acacia species in Africa and the Americas have enlarged, swollen thorns in which ants make their homes. Only limited study has so far been done of the African kinds. But in the 1960s, a scientist named Dr. Daniel Janzen investigated the relationship between New World acacias and their ants. These American species are often called "bull's-horn" acacias, because the large, paired thorns sticking out from the stem resemble very much the horns of a bull.

Eleven species of acacia in the American tropics have consistent relationships with ants. All the ants are closely related, but there are at least ten separate species of them. Not only do the acacias provide the ants with a place to live, they also provide almost all their food. Bull's-horn acacias have very abundant nectaries outside of the flowers, and the ants collect freely from these. The most important and interesting sources of food for the ants, however, are the Beltian bodies, named for Belt, who discovered them.

The leaves of acacias are small and finely divided. At the tips of bull's-horn acacia leaves grow oval or tear-shaped extensions. Instead of being green like the rest of the leaf, ripe Beltian bodies are orange. The ants rely on them for most of their food, and worker ants go out regularly to collect them. They are cut up and fed to the larvae. The Beltian bodies serve no other function for the plant, for during the rainy season, when the acacia is growing too fast for the ants to harvest all of them, the uncollected bodies drop off or rot. During the dry season, when most acacias lose their leaves, bull's-horn acacias retain theirs, and the ants are not deprived of their main food source.

The thorns provide a fine nesting place for the ants. While the thorn is still green the ants cut an entrance hole in it and clear out the small amount of soft pulp inside. The outside of a mature thorn is hard and water-

The "horns" on the bull's-horn acacia make excellent ant-colony homes where, as shown in the cutaway drawing, the young can be reared in safety.

resistant, while the inside walls are absorbent. They probably soak up any excess fluids which might build up in the nest. The hard outer wall helps keep moisture in during the dry season and keeps the rain out during the rainy season. The hole through which the ants go in and out of the thorn is usually blocked by the head of a worker, so the thorn is essentially waterproof.

One colony of ants will occupy one acacia plant. Large colonies may spread out to colonize a few other plants as well. Most of the thorns are occupied by ant brood. The queen lives in one of the larger thorns. As the acacia grows, she moves from one large thorn to another higher up, close to the center of the bush.

Dr. Janzen, by carefully observing the behavior of the ants and by performing experiments, proved without a doubt that the ants protect the acacias from enemies. They provide several services to the plants. Many of them that run about on the branches and leaves of the bush are not on their way to or from a particular task. They are patrolling, on the lookout for enemies. If an insect should be unfortunate enough to crawl onto an acacia, the ants will attack it, stinging and biting until it is stretched and torn to pieces. The remains are usually tossed out of the bush. Small caterpillars are often stung and paralyzed, then taken to a thorn, most likely as food for the ant larvae.

The patrolling workers are also very sensitive to vibrations and to movement near their plant. A large animal moving a yard away causes a strong alarm re-action in the ants. They release an alarm chemical which excites other workers nearby. The aroma is strong enough that a person can smell it. The scent can carry downwind to another acacia ten yards away and disturb workers there as well. Merely waving a finger four inches from workers will alarm them, and shaking the bush excites them greatly. The ants have a strong and painful sting, and people are reluctant to cut down acacia trees for fear of the ferocious attack of the insects.

The ants do other things for the acacia, too. They remove any damaged parts of the plant and clean the bark, leaves, and thorns by licking them. The plants always have a freshly washed look. The ants pay special attention to new leaves. They tend to stick together, but the ants lick away the sticky material and help to separate the new leaflets. Any dust or dead leaves of other plants which may fall on an occupied acacia are removed by the ants. They also chew away any vine or other plant which comes in contact with their home, and they clear an area of ground around the base of the bush so that no other plants grow there. These actions keep the acacia free from competing plants. It has a clear area of access to sunlight all around it.

Janzen proved how important all these ant activities are to the acacias by removing ant colonies from some plants and seeing how they survived. He found that they became severely damaged by insects and that the area around them soon was populated with other plants, blocking out sun from the acacia. These antless plants rapidly declined in health. He estimated that a bull's-horn acacia could not survive more than a year without a colony of ants to protect it.

The acacias have become completely dependent on ants for survival. Their leaves lack the poisons present in at least some other acacias, and have a mild flavor, to human taste at least. Since the new leaves of an unoccupied acacia tend to stick together, they need the ants there to lick them and help them separate.

In especially dry and harsh areas, a different kind of ant is found living in acacias. Although they are closely related to other acacia ants, they have a very different colony structure. Their colonies are probably the largest in the world, for a single one may occupy hundreds of acacia trees. Instead of having only one queen, such colonies may have tens of thousands of them. Normally, when an ant colony produces winged kings and queens, these insects fly off before mating. The queens form their new colonies away from the parent nest. In this kind of acacia ant, however, the young queens mate on or near the parent colony's acacia. Some of them are accepted right back by the colony instead of being driven away.

Such a large colony with many queens has a better chance of survival in difficult circumstances. If the single queen of a normal ant colony dies, the colony dies as well. But if there are many queens, the loss of one is not very important. If the colony occupies many trees, it can easily regroup if one tree loses its leaves or dies due to drought.

The mutualism between ants and acacias has existed long enough for other animals to adapt to it. Two kinds of caterpillars somehow are able to feed on occupied acacias without being attacked by worker ants, and another caterpillar seems to be completely immune to the effects of ant bites and stings, for it totally ignores ants which attack it. Other insects can sense that worker ants are coming to attack them and jump off the plant before becoming victims. Another approach has been used by a weevil, a spider, and a fly. Through evolution, these creatures have come to look and behave so much like the acacia ants that they can live on the bushes with them in complete peace, protected from enemies by their unsuspecting hosts.

One ant relative of acacia ants takes advantage of the ant-plant mutualism in a "sneaky" way. These ants live in thorns and harvest nectar and Beltian bodies, but run and hide if an intruder comes along. They do not protect or help the acacia in any way, so they are truly parasites on the system. Instead of building up a colony with many workers which can protect the acacia, these ants produce winged reproductives within about two months of founding a colony. Thus they take over an acacia temporarily and reproduce quickly before the acacia dies from lack of care. In contrast, true acacia ant colonies put their energies into producing workers for the first two years, using as many as one-third of the worker force to patrol the plant. If the workers of a true acacia ant species find a colony of the parasitic ants, they will attack them and force them out of the acacia.

More Ant Plants

Other relationships between ants and plants are just as interesting as the bull's-horn acacias and their ants. Like the acacias, these plants were described many years ago, but only recently has study of their relationships with ants begun.

Barteria is a small tree which lives in rain forests of Africa. This plant has hollow branches which stretch straight out from the trunk, making ideal ant homes. Young queens of a particular ant species enter the branches and begin their families. When the ant colony is about a year old there are enough workers to patrol the whole tree. While acacia ants seem to protect their plants chiefly from insects, the Barteria ants probably evolved mainly as sentinels against browsing mammals. They are especially good at locating large objects, and their sting is powerful and deep. It causes deep muscular pain in a human for two days. These ants are several times the size of acacia ants and have a habit of jumping down off the branches of their tree frequently. This produces an almost constant "rain" of ants, which sting any animal they land on. The leaves of Barteria are apparently enjoyed by various animals, for one scientist watched while a black colobus monkey ate half the leaves of an unoccupied plant in a forty-minute period. Three branches were broken by the monkey as well. The tree had been occupied by ants before, but with such treatment by hungry animals it probably had only a short time to live now without ants to protect it.

Barteria does not provide food for its ants, only a nesting site. But the protected hollows of the branches are an ideal place for the ants to tend the scale insects and fungus gardens they use for food. So the tree is providing them with prime "agricultural land" and a safe home in return for protection. The ants also clear the area beneath the tree of other plants and clean the leaves.

Other New World plants besides the acacia harbor ant protectors. The Cecropia plant is not able to reproduce if it is heavily shaded. But fortunately, Cecropias are usually occupied by ants which busily chew away any vine which grows close to the trunk. Janzen took vines and twined them around Cecropia trunks. Within eight days the ants had chewed three-fourths of the vine tips to death, and most of the others had also been attacked. Cecropia has hollow branches that ants live in and also provides food. Bits of a very nutritious orange material called Mullerian bodies are produced by the plants. They are embedded in hairy mats of tissue and are harvested by the ants. The ants also tend scale insects inside the branches. But these do not provide honeydew. They are more like beef cattle than dairy cows, for the ants eat the insects themselves.

One of the most amazing associations of ants with plants occurs in southeast Asia. Although these ants and plants have been known since late in the nineteenth

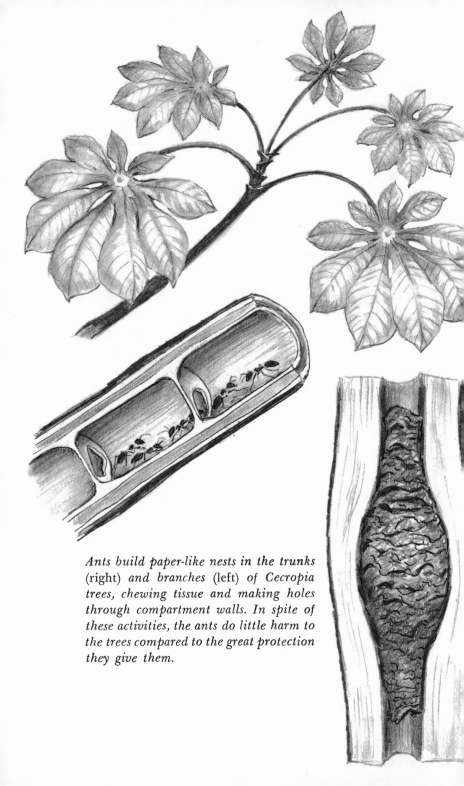

Ants build paper-like nests in the trunks (right) *and* branches (left) *of Cecropia trees, chewing tissue and making holes through compartment walls. In spite of these activities, the ants do little harm to the trees compared to the great protection they give them.*

century, it was only in 1974 that Janzen studied them sufficiently to learn something of the nature of their association. Until then there was much uncertainty about the degree to which the ants and plants benefited one another.

We usually think of plants as having roots in the ground which draw nutrients from the soil. But there are many small plants, called epiphytes, which live entirely above the ground. They are anchored by roots to the trunks or branches of trees and gain nutrients in various ways. Epiphytes are commonly found in tropical climates where rainfall is abundant.

On the island of Sarawak, Janzen found four epiphytes living in close association with one species of ant. One plant is a fern, another a vine, and two are plants with tubers, or thickened stems. One of them is quite common. A stem with a few leaves grows up from the tuber, and short, thick roots connect one side of it to a tree trunk or branch. Inside, the tuber is riddled with natural cavities which house ant colonies. Large holes on the side connecting the tuber to the tree provide entrances for the ants, and scattered small holes give ventilation to the nest. The ants do not make the cavities or holes; they are produced through the growth of the tuber.

The insects do not protect the plant from being eaten, so scientists have questioned what benefit the plant could possibly get from the housing arrangement. Janzen found that these ants differ in one vital way from other tree-dwelling species. Instead of tossing their "garbage" out of their nests, they place it inside the nest. The cavities of the tuber are not all the same. Some of the walls are smooth and hard while others are bumpy. The smooth walls are water-resistant, while the rough walls are not. The ants raise their brood, or young, in the dry, smooth-walled cavities, and they deposit debris in the rough-walled areas. The bumps on the walls apparently act like roots and absorb nutrients from the decomposing ant garbage, thus providing food for the plant.

The ants also tend scale insects on the tree branches, obtaining honeydew from them. They build a protective cover over them from a material called "carton," which the ants make from dirt and debris. The ants collect the seeds of the tuber plant and place them in the carton. There the seeds germinate and grow, providing the ant colony with more space to expand its nest.

The second kind of tuber plant is similar to the first, but it is rare. The fern is rare, too. It looks nothing like familiar ferns, for it consists mostly of a dense mat of branching rhizomes, or special stems that in this case are useful to ants. The rhizomes have cavities inside with a dark lining, making them suitable for the ants' nest. The ants pack their refuse in side branches of the cavity, near where the few fern leaves grow. As these fill up, the ants put debris in the main cavity as well. The rhizomes can soak up nutrients from the decomposing garbage. The ant brood is raised in other parts of the rhizome cavity. The leaves of the fern produce spores rich with oil, and the ants harvest these at night and eat them.

Perhaps the strangest of the four ant plants in this group is the vine. In addition to normal leaves, this vine produces special ant leaves. The ant leaves are like large leaves which have been rolled up to form a cavity inside them. There is a small entrance hole left at one end of the leaf. A root grows from the stem region into this entrance to occupy part of the leaf cavity. The ants also nest in the leaves, which have dark, dry inside walls.

First the adults put brood in the ant leaf. Later they place their refuse there, too, and gradually the leaf is converted completely into a garbage dump. The root which grows into an unoccupied leaf or a leaf which contains only brood is short. As debris is added, the root also grows and branches out until the whole leaf cavity is stuffed with tangled roots. Apparently the ants' garbage makes very good compost for these plants to "feed" on. The ants also collect the vine seeds and plant them in the protective covering of their scale insects.

One ant colony may occupy any combination of these four ant plants, but Janzen never found nests of this ant anywhere else. One colony may have over 10,000 workers and be spread out over many ant plants in several nearby trees. All four of the ant plants provide dark, dry nest cavities, and the ants provide food to all the plants. Only one of the plants gives food in return, and none of them need protection from plant-eaters. The vine releases large quantities of rubbery sap when cut, and the other plants have bad-tasting leaves, at least to humans. So it seems that in this case the ants and the plants are making a simple trade—shelter for the ants in exchange for food for the plants.

6

Galls and Insect Parasites

We have seen how some plants and insects live together in helpful harmony and how others are constantly developing new weapons in their evolutionary arms race. Now we will consider some insects which seem far ahead of the game. They exploit plants to the full, and the plants seem to have little defense in return.

Galls are abnormal growths of plants, similar to tumors. One rather often sees galls while walking through the woods. Some of them are merely small bumps on leaves. Others are misshapen lumpy growths on twigs or stems, while still others are smooth bulges. Some galls are as big as a baseball. They may be caused by bacteria, fungi, roundworms, mites, or insects. Galls are formed when a plant parasite attacks and the plant reacts with extra growth around the attacker.

Although the plant's reaction may be defensive, it results in protection of the parasite rather than injury to it. A cozy, protected home is produced. Perhaps the

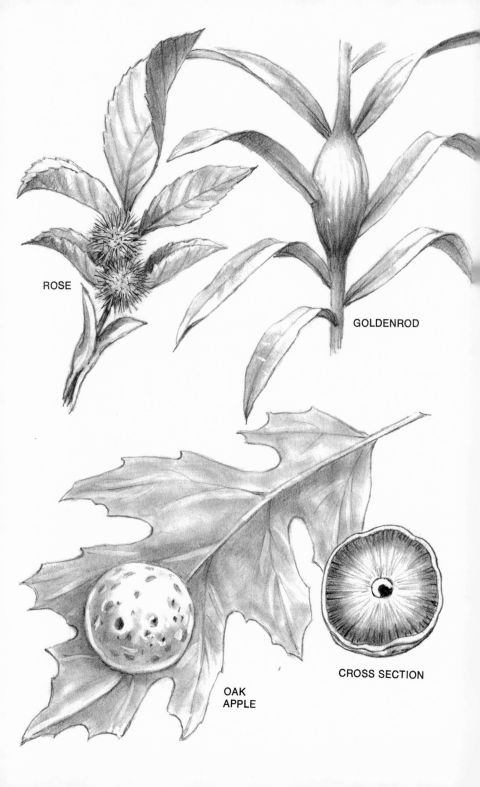

ROSE

GOLDENROD

OAK
APPLE

CROSS SECTION

plant gains something, however, since at least the parasite is forced to remain localized in one part of it. In the case of the fig wasp, the wasp from the gall pollinates the plant, so the plant gains something. But other gall insects are pure parasites, taking advantage of the growth reaction that gives them food and a protected place to live.

Although galls may form on any plant part, each kind of gall insect attacks a specific site. The attack results in a characteristic reaction on the part of the plant. The complexity of galls is amazing. Structures develop which the plant does not otherwise form. Cavities and galleries surrounded by hard protective coverings are often formed. Soft, juicy tissues which the insects feed on are produced. The development of the gall is synchronized with that of the insect. Galls which are completely enclosed while the insects are growing may crack at the proper time, allowing the insects to leave. Galls with hard surfaces sometimes even provide a convenient exit hole with a plug which pops out and lets the insect free.

A great assortment of insects cause gall formation. Many beetles, aphids, wasps, moths, and flies are gall

Left: *Galls come in many varieties, according to the plants used and the kind of insect or other organism causing them. A rose gall is shown at upper left, and a goldenrod gall at upper right. Below, the oak gall, or "oak apple," with a cross section revealing the curled-up larva at the center.*

insects. Most galls begin to form when the larvae start feeding. Some start right after the eggs are laid, while others can be induced by adult insects.

While many kinds of plants can be "tricked" into forming galls, some are more susceptible than others. Almost three-fourths of gall insects in this country parasitize plants from just twelve families. Roses, goldenrod, and willows are among the most frequent victims. But the number-one host is the oak. Wasps alone induce 591 different kinds of galls on oaks. Almost half of these are leaf galls, but others are found on stems, roots, buds, leaves, flowers, and acorns.

One gall-forming weevil specializes in cabbage roots. The adult feeds on flower buds. Mating takes place in the spring. Then the female goes to the ground. Using her proboscis, she scoops a cavity in the root and lays her egg there. The root, in healing, obligingly grows over the egg, protecting it. When the larva hatches and begins to feed, the root swells up. After four weeks the larva bores out and pupates in the soil.

Some gall insects have a very complicated life cycle. They do not have only ordinary generations, with males and females which mate; they also have a purely female generation. These females can lay eggs without having mated. Some cynipids, relatives of bees and wasps, have this system. The offspring of the two generations inhabit very different galls. The two types of females and their galls are so different that scientists often assigned them to totally different species.

One cynipid parasitizes oak roots and buds. Males and females emerge in July from the bud gall and mate. The female, which cannot fly, goes to the soil and lays her eggs on roots. The small, rounded galls may occur in bunches on the roots. Development is slow. In the winter of the second year the purely female generation emerges. These females are completely wingless and have huge abdomens full of eggs. They climb up to the dormant buds of the oak. Using her long ovipositor, the female pierces right to the growing point of the bud and lays several eggs. The bud gall which develops becomes almost two inches in diameter. It is very complicated in structure, with many rounded lobes and chambers. Plenty of growing tissue is provided by the oak, and larval feeding is no problem. The larvae pupate in the gall and emerge in July, starting the cycle over again. Two years pass from the mating of the male and female until the production of the next sexually reproducing generation.

The Cause of Galls

What causes the formation of galls? This question is yet to be answered and probably has no simple answer. The feeding of the larvae stimulates galls to grow, but some begin growing before the larvae hatch. A study was begun on a leaf-gall sawfly to determine the causes of gall growth, but unfortunately the research was never finished.

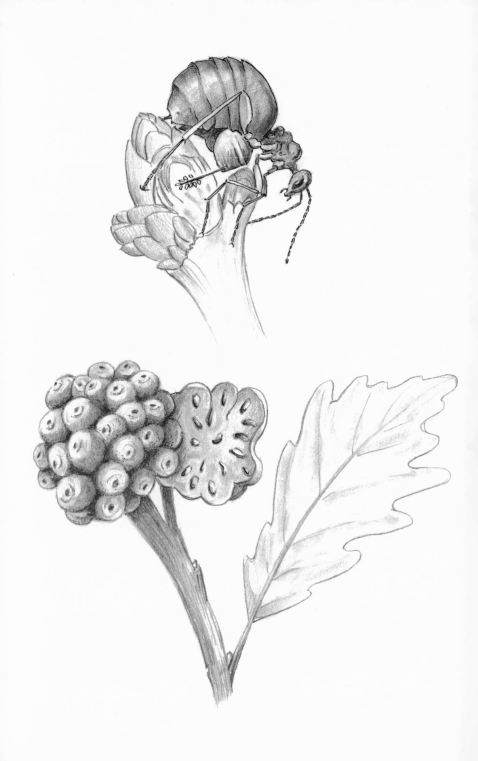

The female of this wasp, attracted by chemical factors of the willow, saws into the midrib of a leaf with her ovipositor. She picks a spot on the underside of the leaf, close to the growing tip. The egg must be placed precisely between the cell layers of the leaf. When she lays the egg, she also injects a fluid which causes the gall to begin. Even if no egg is laid, the first stages of gall growth occur. After four or five days the larva hatches. It remains inside the gall, feeding on the mass of tissue which the leaf produces inside the gall.

Some 35 to 40 days after hatching, the larva pupates. About two weeks later the adult wasp emerges. The female is an egg-laying machine. She lives only a couple of days and does not eat. Her abdomen is swollen with the 50 to 100 eggs which she will lay and with the fluid she will inject along with them.

Scientists found that this fluid causes the early stages of gall growth. Even if no egg is laid, or if the egg is killed, the gall begins to form. But later growth of the gall requires feeding by the larva. More growth occurs on the side where the larva feeds most. The larva moves around, feeding alternately in different parts of the gall, with the result that the growth evens out. The gall almost develops a life of its own. Even if the

Left: *This insect,* Biorrhiza pallida, *is one of the gall wasps. She is a member of a purely female generation and is wingless. She lays her eggs in oak buds, which then develop galls, shown in normal form and in cross section below.*

leaf withers and falls from the tree, the gall stays green. It may even put out roots as if attempting to remain alive no matter what.

The Well-Adapted Aphids

Aphids as a group are certainly the most successful plant-feeding insects. They belong to the same insect family as scale insects, cicadas, and water bugs. This strange assortment has in common a mouth with piercing, sucking mouth parts. The water bugs use their mouths to suck out the juices of other water animals. Cicadas use theirs in the larval stage to suck root juices from trees. Scale insects and aphids suck various plant juices.

The success of aphids is based on their feeding specializations. More that 99 per cent of aphids have a strong preference for just a few kinds of host plants, while more than half parasitize just one plant species. Many plants seem to be immune to aphids, since only about 10 per cent of the higher plants have aphids which choose them. Some aphids prosper by alternating between a woody winter host and a juicier, annual summer host.

Settling Down

Aphids can be very choosy about picking a spot to feed. After landing on a plant an aphid will poke its sharp feeding stylets into it, apparently sampling its chemical makeup. If the site is unsuitable, the aphid will fly off and land again elsewhere. Many aphids are stimulated to feed by the type of sugar called sucrose. Little or no feeding will occur if no sucrose is present. The preference for sucrose varies, however. The pea aphid requires a high sugar concentration, while the cotton aphid does not. The cabbage aphid uses sinigrin as a specific host-selection clue. Other kinds of aphids prefer other chemicals or combinations of them.

Once it has selected a feeding site, the aphid settles down and inserts its stylets in the food-conducting tissues of the plant. The pressure of the sap is apparently high enough that the aphid does not need to suck actively. It just sits there and lets the sap flow into its body. Scientists have taken advantage of this fact in their studies of plant sap. They allow aphids to establish themselves on experimental plants. Then they cut off the aphids' bodies, leaving the stylets inserted in the plant. Drops of pure, unaltered sap then come from the stylets and the scientists can analyze them.

As we have seen, ants take advantage of the abundance of food received by aphids, too. More liquid and sugar passes into the aphid than it can use. So it passes the sugary honeydew out its other end. Another kind of sucking aphid relative also produces honeydew. One species in the Middle East produces glistening scales of honeydew when it feeds on the tamarisk tree. These scales float to the ground and are used by nomadic tribes for food. The "manna from heaven" which fed the Israelites while they were wandering in the desert thousands of years ago is thought to have been this flaky honeydew.

Even when not attended by ants, aphids thrive. Like some other insects, they produce females which need not mate in order to bear young. They produce living young instead of laying eggs. All summer they reproduce this way. A female may give birth to a dozen young in 24 hours. Reproduction proceeds so fast that embryos begin to develop inside the larvae before they are even born. As the day length shortens in the fall, however, a generation of males and females is produced. After mating, the females lay the overwintering eggs. These hatch in spring into females only. More males are not produced again until fall.

When they are crowded, aphids produce winged females. In contrast to the sluggish-feeding usual aphids, these insects are restless and fly off. They are attracted by the ultraviolet light of the sky. After flying a short way, their orientation changes. Now they are attracted to green foliage and alight on the nearest plant. If it is not suitable, they fly off again. If it is, they settle down. Their wing muscles dissolve and provide nutriment to the developing young. The restless flier has become the settled feeder, and moves around very little for the rest of her life.

Aphid Galls

Many aphids induce gall-formation. Their galls are the most specialized types produced by insects. A good example is the poplar gall. A female aphid, hatched from an overwintering egg, settles on the tender young leaf stem of a poplar. As she sucks the tree juices, plant cells near her are stimulated to grow. Eventually they grow over her body, enclosing her in a cavity with a small opening. The aphid molts a month or so after establishing herself and produces a batch of 20 to 30 wingless young. These grow and then reproduce, forming larvae which molt into winged adults. It gets pretty crowded inside the gall, which is less than an inch long. In August the winged females fly off and lay eggs in crevices of poplar bark. Males and females hatch from these eggs. After they have molted and grown to maturity, they mate. These mated females then lay the overwintering eggs.

7

The Insect-Eating Plants

Most of this book has concerned ways insects feed on plants and plants defend themselves in return. There are a few plants, however, which have turned the tables and feed on insects.

Most insect-eating plants are found in soggy, acid soil. Minerals are scarce in these locations, and scientists think that insect-eating evolved as a way of obtaining the nitrogen and phosphorus necessary for vigorous growth. The trapped and digested insects are a kind of fertilizer, absorbed through the leaves rather than the roots. Roots of some carnivorous plants are rather poorly developed, probably because they do little to help feed the plant.

Carnivores are found in several plant families. Their traps range in size from the microscopic worm-catching nooses of some fungi to foot-deep urns holding two quarts of water in which small mammals as well as insects may drown. Different methods are used to lure and trap the insects. In addition to nooses and pools, carnivorous plants use sticky secretions and snapping traps to capture their prey.

The Venus's-flytrap is a very remarkable adaptation in the plant world; it is literally a trap, and easily snapped shut by an insect's movements.

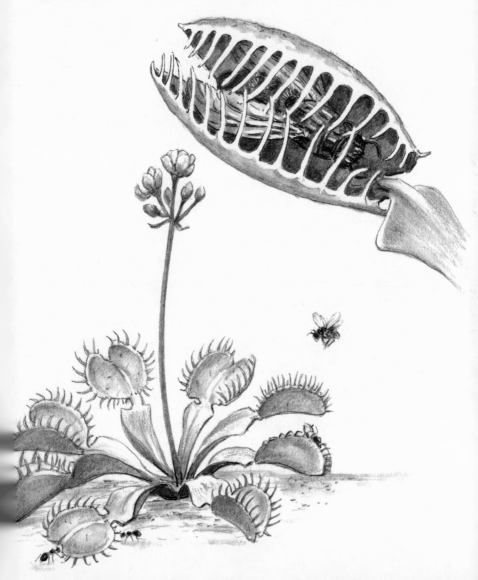

Venus's-Flytrap

Most famous of all these plants is the Venus's-flytrap. Actually, this plant is one of the rarest carnivorous species. It is found only in acid peat bogs along the Carolina coast. It is in danger of extinction because of its restricted range and its popularity as a "pet."

The plant is small, rarely more than six inches across. The traps are found at the ends of the leaves. Each trap has two halves more or less suggesting teaspoons, and edged with sharp spines. On the inner face of each half are three small trigger hairs. If two of these are touched, the trap snaps shut. The spines mesh, creating a cage from which only very small insects can escape. Glands on the inside, which color the trap red, secrete digestive juices which dissolve the prey completely in a few days. Although a Venus's-flytrap can survive without feeding, fed plants grow and thrive. Scientists do not know for sure how the trap can close so fast since plants lack nerves and muscles. They feel there is some kind of tension in the open leaf which is released when the hairs are touched.

Deadly Dewdrops

Although the plants look very different, the Venus's-flytrap is a close relative of the sundews. It is amazing that two such different trapping techniques should evolve in the same family. While the flytrap is a single, localized species, sundews have more than 90 species found around the world.

The familiar eastern American sundew is only about an inch across. It has small paddle-shaped leaves covered with minute hairs. The rounded, red tip of each hair is engulfed in a drop of liquid which sparkles in the sun. Other sundews have long, straight leaves which point upward and are covered along their whole length by the glandular hairs. Some have forked leaves and some have flattened leaves. In Australia, sundews are found with yard-long stems.

If an insect touches one of the droplets on a sundew hair, it tries to struggle away. Smaller insects are not strong enough. If anything, they become more entangled with other drops. The victim's struggles stimulate nearby hairs to bend over, slowly engulfing the prey. Digestive enzymes in the secretion break down the insect tissues, and the nutrients are absorbed by the hairs into the body of the plant.

Moths are immune to sundews, for the scales on their wings keep them from being engulfed. They simply bounce off and fly away, leaving a few scales behind. Perhaps because the adults could approach without danger, the larvae of a small but remarkable moth feed on sundews. The larvae come out at night. Instead of avoiding the deadly hairs, one of these hungry caterpillars rears up and drinks the sticky fluid. It then chews down the whole hair and repeats

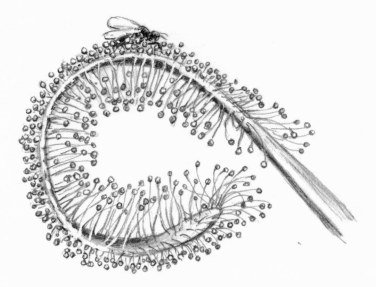

The sundew traps insects in a different way from Venus's-flytrap—with a biological "glue." Most insects quickly become stuck on it, but the caterpillars of one species of moth, below, can safely drink the glue and eat the hairs that produce it.

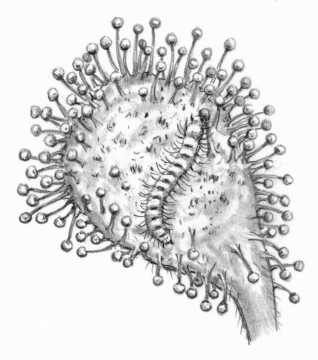

the performance again and again, until the whole center of the leaf has no more hairs. The caterpillar leaves hairs around the leaf edges, however. Perhaps this helps keep away possible enemies. The caterpillar will even eat the partially digested remains of less fortunate insects trapped by the plant.

Death by Drowning

Pitcher plants belong to several species in different parts of the world. Several are found in the United States. All have one thing in common—a tube-shaped leaf with fluid inside. Some pitchers flare outward and collect rain with ease. Others have "umbrellas" or hoods over them that keep out the rain. Some lie down instead of standing up straight.

A typical kind of pitcher plant is found all along the eastern coast of the United States and north into Canada as far as Labrador. It has an 18-inch banana-shaped hollow leaf. The edge of the leaf is scalloped. It secretes nectar which attracts insects. But once they land, they find themselves standing on a bed of sharp spines which point inward toward the water-filled interior. They cannot climb up and away from the fluid because of the spines. So they climb downward, until they reach a waxy, smooth zone where they slip and fall, ending up floating dead in the pitcher fluid. Although mostly water, the liquid also contains digestive enzymes and bacteria which aid in digestion.

Another American pitcher plant has horizontal leaves. Rainwater cannot enter because of the hood over the leaf entrance. But even after a prolonged drought, a small amount of liquid containing insect victims is found at the bottom of the tube. In this species, long flexible hairs pointing toward the base of the leaf grow throughout the length of the tube. They are so long that they meet in the center. Once a small insect has entered the leaf, it can go in only one direction, toward its death at the bottom.

The tropical plant Nepenthes has pitchers which develop from tendrils extending out from the thick, glossy, yard-long leaves. Nepenthes has a lid which helps keep out rainwater. Around the top edge of the pitcher are glands which secrete a highly attractive, sugary liquid. Many insects are lured to the pitcher, and even wasps and cockroaches are found dead inside. The secret of Nepenthes lies not in inward-pointing spines, but in loose, waxy scales. An insect trying to escape climbs up the pitcher walls with great difficulty. It wipes its feet together and onto its body. But eventually it fails completely and slides into the digestive fluid at the bottom.

The inner pitcher surface feels slimy to the touch. But with the electron microscope, minute, overlapping

Right: *Pitcher plants have leaves forming a vase at their base; secreted nectar attracts insects that become trapped. In Nepenthes, at left, insects slip down on a waxy surface; in Sarracenia, below, insects are trapped by hairs, as shown in the cutaway enlarged leaf.*

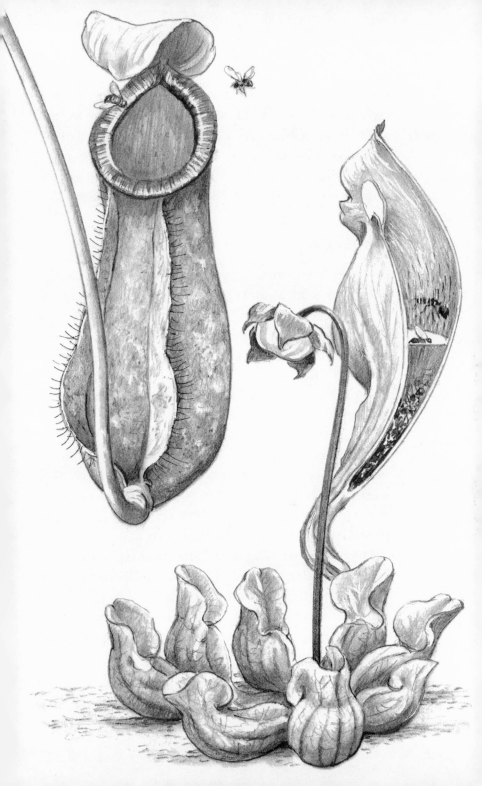

waxy scales can be seen. Under the layer of scales is a layer of waxy ridges. These ridges make it easy for the scales to detach and stick to insects' feet. The harder an insect tries to escape from Nepenthes, the more scales collect on its feet until it is completely unable to hold on and falls into the deadly liquid below.

Other Carnivores

Other carnivorous plants exist. Some, such as the butterwort, have sticky surfaces which trap insects. Some water plants are meat-eaters. One kind is a close relative of the Venus's-flytrap. It catches water bugs, water fleas (a type of crustacean), and other small aquatic animals. The bladderworts are smaller and prey on protozoa, water fleas, worms, and insect larvae.

The smallest carnivore of all is the strange microscopic fungus which traps small roundworms in its nooses. Each noose consists of just three cells. If a worm crawls into the noose, all three cells swell up at once, trapping the worm. Projections of the fungus penetrate the worm's body and digest it.

Over the ages, plants and insects have evolved an excellent balance between them. Both are able to thrive, and they use one another in many different ways to better their lives. As scientists study them more, we are surely in for further surprises concerning the way insects and plants have adapted to one another.

Suggested Reading

Books

Cooper, Elizabeth, *Insects and Plants, The Amazing Partnership* (Harcourt, N.Y., 1963)
Goldstein, Philip, *Animals and Plants That Trap* (Holiday House, N.Y., 1974)
Hutchins, Ross E., *Strange Plants and Their Ways* (Rand McNally, N.Y., 1958)
———, *Galls and Gall Insects* (Dodd, N.Y., 1969)
Martin, Lynne, *The Orchid Family* (Morrow, N.Y., 1974)
Mason, Herbert M., Jr., *The Fantastic World of Ants* (McKay, N.Y., 1975)
Meeuse, J. D., *The Story of Pollination* (Ronald Press, N.Y., 1961)
Schwartz, Randall, *Carnivorous Plants: Their Care and Feeding* (Praeger, N.Y., 1974)
Simon, Hilda, *Insect Masquerades* (Viking, N.Y., 1968)
———, *Our Six-legged Friends and Allies: Ecology in Your Back Yard* (Vanguard, N.Y., 1972)
Waters, John F., *Carnivorous Plants* (Watts, N.Y., 1974)

Magazine Articles

Arditti, Joseph, "Orchids," *Scientific American,* Jan. 1966 (Offprint #1031)

Ashley, Terry, and Joseph F. Gennaro, J., "Fly in the Sundew," *Natural History,* Dec. 1971

Batra, Suzanne W. T., and L. R. Batra, "The Fungus Gardens of Insects," *Scientific American,* Nov. 1967 (Offprint #1086)

Brower, Lincoln P., "Ecological Chemistry," *Scientific American,* Feb. 1969 (Offprint #1133)

Davidson, Treat, "Rose Aphids," *National Geographic,* June, 1961

Ehrlich, Paul, and Peter H. Raven, "Butterflies and Plants," *Scientific American,* June 1967 (Offprint #1076)

Eisner, Thomas, "Life on the Sticky Sundew," *Natural History,* June 1967

Grant, Verne, "The Fertilization of Flowers," *Scientific American,* June 1951 (Offprint #12)

Hovanitz, William, "Insects and Plant Galls," *Scientific American,* Nov. 1959

Moser, John C., "Trails of the Leafcutters," *Natural History,* Jan. 1967

Pramer, David, and Norman Dondero, "Microscopic Traps," *Natural History,* Dec. 1957

Rothschild, Miriam, and Bob Ford, "Heart Poisons and the Monarch," *Natural History,* April 1970

Sisson, Robert F., "The Wasp That Plays Cupid to a Fig," *National Geographic,* Nov. 1970

Zahl, Paul A., "Plants That Eat Insects," *National Geographic,* May 1961

———, "Malaysia's Giant Flowers," *National Geographic,* May 1964

Index